Springer Theses

Recognizing Outstanding Ph.D. Research

Aims and Scope

The series "Springer Theses" brings together a selection of the very best Ph.D. theses from around the world and across the physical sciences. Nominated and endorsed by two recognized specialists, each published volume has been selected for its scientific excellence and the high impact of its contents for the pertinent field of research. For greater accessibility to non-specialists, the published versions include an extended introduction, as well as a foreword by the student's supervisor explaining the special relevance of the work for the field. As a whole, the series will provide a valuable resource both for newcomers to the research fields described, and for other scientists seeking detailed background information on special questions. Finally, it provides an accredited documentation of the valuable contributions made by today's younger generation of scientists.

Theses are accepted into the series by invited nomination only and must fulfill all of the following criteria

- They must be written in good English.
- The topic should fall within the confines of Chemistry, Physics, Earth Sciences, Engineering and related interdisciplinary fields such as Materials, Nanoscience, Chemical Engineering, Complex Systems and Biophysics.
- The work reported in the thesis must represent a significant scientific advance.
- If the thesis includes previously published material, permission to reproduce this must be gained from the respective copyright holder.
- They must have been examined and passed during the 12 months prior to nomination.
- Each thesis should include a foreword by the supervisor outlining the significance of its content.
- The theses should have a clearly defined structure including an introduction accessible to scientists not expert in that particular field.

More information about this series at http://www.springer.com/series/8790

Adrian A. Valverde

Precision Measurements to Test the Standard Model and for Explosive Nuclear Astrophysics

Doctoral Thesis accepted by the
University of Notre Dame, Indiana,
USA

 Springer

Adrian A. Valverde
University of Manitoba
Winnipeg, MB, Canada

ISSN 2190-5053   ISSN 2190-5061   (electronic)
Springer Theses
ISBN 978-3-030-30780-6   ISBN 978-3-030-30778-3   (eBook)
https://doi.org/10.1007/978-3-030-30778-3

This Springer imprint is published by the registered company Springer Nature Switzerland AG.
The registered company address is: Gewerbestrasse 11, 6330 Cham, Switzerland

Supervisor's Foreword

Precision measurements of ground state properties of atomic nuclei are critical to better understanding of our Universe from the infinitesimally small to the astronomically large. Indeed, physics beyond the Standard Model (SM) of elementary particles can be constrained through a variety of probes of nuclear properties including the unitarity test of the Cabibbo–Kobayashi–Maskawa (CKM) matrix. This rotation matrix takes the regular mass Eigenstates of quarks and turn them into Eigenstates under the weak interaction. A non-unitarity of this matrix could have deep consequences for the SM, including new physics such as a missing quark generation, new bosons, supersymmetry, or it could be the result of systematic effects not fully understood yet. The most precise test of the CKM matrix unitarity currently comes from the normalization of the top row, where the largest element, V_{ud}, is most precisely calculated through the determination of a series of comparative half-lives, or ft-values, of superallowed pure Fermi beta decays. The accuracy of this value can be tested by determining V_{ud} from other superallowed transitions such as the transitions between mirror nuclei. One of the four experimental quantities requiring a precise measurement to do this is the half-life of the decaying nuclei.

While the half-lives of several mirror transitions has been improved by various groups around the world in the past few years, the longest-lived remained untouched, including the decay of ^{11}C to ^{11}B, with a half-life of approximately 20 min. This long-lived mirror nuclei also happens to be lighter than the measured nuclei and thus its ft-values would be more sensitive to the presence of scalar currents, one possible form of beyond the SM physics, in the decay. Hence, based on these considerations, Dr. Adrian Valverde performed the most precise measurement of the ^{11}C half-life to date. This measurement was accomplished using radioactive ion beams (RIB) from the TwinSol facility of the University of Notre Dame's Nuclear Science Laboratory. A total of 9 decay curves, each measured over periods of almost 7 h and with different initial beam rates, and experimental settings to investigate possible systematics, resulted in a half-life of 1220.27(26) s, which is over 5 times more precise than the world average. This measurement then resulted in the most precisely known ft-value of all mirror transitions.

Precision measurements of another ground state property of a nucleus, its atomic mass, can be used to better understand the physics of explosive astronomical events such as X-ray bursts, where the envelope of a large star gets aspired by a neutron star resulting in a thermonuclear explosion, or the merger of two neutron stars, resulting in another explosive event, a kilonova. Both types of events are fueled by a series of nuclear reactions. In the former, these nuclear reactions are a series of rapid proton captures followed by beta decays, called the *rp*-process, while in the latter they are a series of neutron captures followed by beta decays, called the *r*-process. In both cases, the relative abundance of neighboring nuclei at equilibrium during the nucleon capture process is given by a detailed balance equation that depend exponentially on the mass differences between these nuclei. Hence, the abundance of nuclei produced by these nucleon-capture processes are extremely sensitive to atomic masses and typically relative uncertainties on the order of 10^{-6} or 10^{-7} are required.

The second part of Dr. Valverde's thesis consists of measuring, for the first time, the atomic mass of ^{56}Cu. Prior to that measurement, network calculations for X-ray burst events had to rely on an atomic mass either derived from local mass models or from the extrapolation of existing masses in the region. These different values resulted in significant differences in the flow pattern around the waiting point ^{56}Ni. As the name implies, these significantly longer-lived nuclei puncturing the *rp*-process can severally slow down the process and in the extreme case it could even stall it, preventing reaching heavier masses. Dr. Valverde's mass measurement of ^{56}Cu was performed at the LEBIT Penning trap facility of the National Superconducting Cyclotron Laboratory of Michigan State University. The new atomic mass obtained results in a flow partially bypassing the ^{56}Ni waiting point via the route ^{55}Ni$(p,\gamma)^{56}$Cu. This flow results in an enhancement in the production of heavier masses accompanied by a reduction in the region of ^{56}Ni.

The last part of Dr. Valverde's thesis consisted in constructing a critical component of an upcoming facility that will aim at producing many critical nuclei for the *r*-process happening during a neutron-star merger. This second type of nucleon-capture process is responsible for producing about half of all elements heavier than iron while being even less known than the *rp*-process since it involves very neutron-rich nuclei that for the most part have never been synthesized in the laboratory. While we are starting to get more information on medium-mass nuclei with up to about 65 protons, the situation is worse for heavier nuclei as fission reactions cannot produce them and the production cross-section of fragmentation reactions rapidly drops with neutron richness. This region can however be accessed via multi-nucleon transfer reactions, where neutron-rich stable target and projectile nuclei exchange multiple nuclei allowing for the production of very neutron-rich heavy nuclei. A new radioactive ion beams (RIB) facility, called the $N = 126$ Factory, currently under construction at Argonne National Laboratory will make use of such reactions to produce radioactive nuclei of importance for the *r*-process. In order to form an RIB that can be used by various experiments, the nuclei that are produced at large angles will first have to be stopped in a large-volume gas catcher before being accelerated at low energy. Many types of nuclear physics experiments require the use of pure RIB containing only the species of interest. Hence, the RIB from the gas catcher will have to be cleaned and the state-of-the-art instrument used to remove contaminants with same atomic mass numbers is a multi-reflection time-of-flight mass spectrometer (MR-ToF). Such devices, however, require the production of ion bunches of well-defined energy. To meet this requirement, as part of his thesis, Dr. Valverde assembled and tested a radio-frequency quadrupole cooler and buncher.

Dr. Valverde's thesis is a complete body of work involving a technical development project, two scientific measurements of high impact, and the corresponding data analysis. All the presented work in this thesis have also been published in an abbreviated format. He has played a leading role in all of these aspects. This thesis work will remain of relevance in multiple fields of nuclear physics including fundamental symmetry and nuclear astrophysics as well as serving as a more technical development presenting the RFQ cooler and buncher of the $N = 126$ factory. I am delighted that Dr. Valverde's dissertation has been chosen for publication in the Springer Theses Series.

Hong Kong   Maxime Brodeur

Acknowledgments

A friend suggested I subtitle my thesis, *A Tale of Three Cities*. The suggestion came after I was committed to a title, but it would have been a good acknowledgment of the spatial component to my journey as a graduate student. My dissertation has benefited significantly from the involvement and input of many people across three geographically distinct institutions, all of whom to which I owe my gratitude not only for their role in bringing this work to completion but also for their role in my formation as an experimental physicist.

First, I would like to express my gratitude to my advisor, Maxime Brodeur, for his mentorship, encouragement, and support, which has been critical to the completion of my degree in more ways than I can adequately express. I would also like to thank my committee, Manoel Couder, Grant Matthews, and Carol Tanner, for their guidance.

My first steps on the road to this degree began at the NSCL, and I am grateful to the lab at large for all of their assistance. I am particularly grateful to Georg Bollen for having given me the opportunity to begin my journey in high-precision nuclear measurements and for his encouragement in my continuing it. I am also thankful to Matt Redshaw for his supervision and encouragement of my undergraduate work, and to Ryan Ringle, whose support and mentorship has lasted beyond my time at the NSCL and has had an important role in the formation of my dissertation. I am fortunate for the fantastic team of LEBIT collaborators I have the opportunity to work with. None of the work done at LEBIT would be possible without the N4 gas cell, and I am particularly grateful to Chandana Sumithrarachchi for his patience and guidance as I iterated through explanations of the work they do in various drafts of papers. I am grateful to the LEBIT post-docs and the three generations of students whose time I overlapped with; I'd like to particularly thank Martin Eibach and Kerim Gulyuz for their assistance and willingness to share their knowledge, and my own generation of students, Chris Izzo and Rachel Sandler, for their friendship and support through this process. I am also grateful to Wei Jia Ong for the speed with which she iterated through the reaction rate calculations and X-ray burst models for ^{56}Cu.

My next step brought me to the University of Notre Dame; I am grateful to the NSL and the *TwinSol* collaboration for the welcoming environment they provided and to the Department of Physics at large for the opportunities they offered me. I would like to thank my fellow members of the Brodeur Group, Dan Burdette, James Kelly, Jacob Long, and Patrick O'Malley, for their friendship and for the productive conversations on our increasingly diverging research.

The itinerant part of my time in grad school came to an end at Argonne National Laboratory. I am grateful to my supervisors at ANL, Guy Savard and Jason Clark, for all of their assistance, encouragement, and support. I would like to thank Dan Lascar for all of his assistance in getting the Cooler-Buncher project off the ground, and John Rohrer, Bruce Zabransky, Frank Smagacz, and the ANL Machine Shop for their efforts helping me complete it. I would also like to thank my fellow "xanltrappers" at Argonne: Mary Burkey, Jeff Klimes, Rodney Orford, Jacob Pierce, Dwaipayan Ray, and Louis Varianno, for their friendship, support and conversations that shaped several sections of this dissertation, and the manual labor necessary for taking several hundred pounds of materials delivered to implausible locations other than the loading dock and turning them into a functional device.

Last, but most importantly, I would like to thank my family. My parents, Gilbert and Carmen; my grandmother, Carol Sue; and my siblings, Ariana and André, who have all been invaluable sources of support and encouragement throughout this process. Without them, none of this would have been possible.

Parts of This Thesis Have Been Published in the Following Journal Articles

- A. A. Valverde et al., "High-precision mass measurement of ^{56}Cu and the redirection of the rp-process flow", Phys. Rev. Lett. **120**, 032701 (2018)
- A. A. Valverde et al., "Precision half-life measurement of ^{11}C: the most precise mirror transition $\mathcal{F}t$ value", Phys. Rev. C **97**, 035503 (2018)
- A. A. Valverde et al., "A cooler-buncher for the N=126 factory at Argonne National Laboratory", Nucl. Instr. Meth. Phys. Res. B (2019)

Contents

Chapter 1
Introduction

1.1 The Precision Frontier

Precision measurements in nuclear physics are an important and active avenue of research spanning tests of the Standard Model through the study of the decay of fundamental particles to the study of how the elements of the universe were synthesized. In the search for physics beyond the Standard Model, high precision searches through the study of nuclear β decays [1, 2] form an important part of a threefold approach that also includes efforts at the high energy [3, 4] and high intensity [5] frontiers. Furthermore, precision measurements offer an important avenue for the study of nuclear structure, including shell and subshell structure, pairing, and deformation; for the determination of astrophysical reaction rates for astrophysical nucleosynthesis pathways, including the r- and rp-processes; and for the study of nuclear models, such as the isobaric mass multiplet equation [6].

This dissertation is comprised of three specific applications of precision measurements in nuclear physics. First discussed will be the precision measurement of the ^{11}C half-life for tests of the electroweak sector of the Standard Model through nuclear β decays. Second, Penning trap mass spectrometry will be introduced, and a specific application through the mass measurement of ^{56}Cu for determining the rp-process flow around the ^{56}Ni waiting point will be presented. Finally, an overview of the $N = 126$ factory will be given, which is a new facility being built at the Argonne Tandem Linac Accelerator System to allow for precision measurements of interest for the astrophysical r-process; the focus will be the radiofrequency quadrupole (RFQ) cooler-buncher, a component of this facility.

© Springer Nature Switzerland AG 2019
A. A. Valverde, *Precision Measurements to Test the Standard Model and for Explosive Nuclear Astrophysics*, Springer Theses,
https://doi.org/10.1007/978-3-030-30778-3_1

1.2 Testing the Standard Model with Nuclear Beta Decays

1.2.1 Nuclear Beta Decay

In a nuclear beta decay, an unstable nucleus X of atomic number Z and neutron number N transitions into a more stable nucleus Y of atomic number $Z \pm 1$ and neutron number $N \mp 1$ accompanied by the emission or capture of a β particle $(e^{\mp})$ and an electron neutrino or antineutrino (ν_e or $\overline{\nu_e}$). This occurs either through the decay of a neutron into a proton with the emission of a e^-, called β^- decay (Eq. (1.1)), or through the decay of a proton into a neutron. This can occur either through the emission of a positron, called β^+ decay (Eq. (1.2)), or through the capture of an orbital atomic electron by the proton, which is called electron capture or EC decay (Eq. (1.3)).

$$\ce{^A_Z X_N} \rightarrow \ce{^A_{Z+1} Y_{N-1}} + e^- + \overline{\nu_e} \tag{1.1}$$

$$\ce{^A_Z X_N} \rightarrow \ce{^A_{Z-1} W_{N+1}} + e^+ + \nu_e \tag{1.2}$$

$$\ce{^A_Z X_N} + e^- \rightarrow \ce{^A_{Z-1} W_{N+1}} + \nu_e \tag{1.3}$$

For each of these cases, the energy released in the transition or Q value can be calculated from the masses of the mother and daughter nuclei m_N (or alternatively the masses of the neutral ions, $m = m_N + Zm_e$) and the mass of the electron or positron, m_e; for β^-, β^+, and EC decays, these are [7]:

$$\frac{Q_{\beta^-}}{c^2} = m_N\left(\ce{^A_Z X}\right) - m_N\left(\ce{^A_{Z+1} Y_{N-1}}\right) - m_e = m\left(\ce{^A_Z X}\right) - m\left(\ce{^A_{Z+1} Y_{N-1}}\right) \tag{1.4}$$

$$\frac{Q_{\beta^+}}{c^2} = m_N\left(\ce{^A_Z X}\right) - m_N\left(\ce{^A_{Z-1} W_{N+1}}\right) - m_e = m\left(\ce{^A_Z X}\right) - m\left(\ce{^A_{Z-1} W_{N+1}}\right) - 2m_e \tag{1.5}$$

$$\frac{Q_{EC}}{c^2} = m_N\left(\ce{^A_Z X}\right) + m_e - m_N\left(\ce{^A_{Z-1} W_{N+1}}\right) = m\left(\ce{^A_Z X}\right) - m\left(\ce{^A_{Z-1} W_{N+1}}\right) \tag{1.6}$$

where the masses of the neutrinos and the electron binding energies have been ignored, as they are significantly smaller than the other energies involved.

Enrico Fermi developed a theory of β decay based on Wolfgang Pauli's neutrino hypothesis in 1934 [8]. The critical feature of this theory can be determined from the expression of the transition probability between quasistationary states where the interaction causing the transition is weak compared to the interaction that forms the quasistationary states; as the timescale of a decay is much smaller than the lifetime of states, this is a valid approximation [7]. The transition rate λ can then be calculated by Fermi's Golden Rule [7]

$$\lambda = \frac{2\pi}{\hbar} |M_{f,i}|^2 \rho \left(E_f \right) \tag{1.7}$$

where the matrix element $M_{f,i}$ is the integral of the interaction $\hat{V}$ between the initial and final quasistationary states, encompassing the dynamical information about the interaction, and $\rho(E_f)$ is the density of final states, a phase space factor incorporating the kinematical information. For a nuclear $\beta^{\pm}$ decay, the final state wavefunction must include both the wavefunction of the daughter nucleus ψ_D as well as the electron and neutrino wavefunctions ϕ_e and ϕ_ν, whereas the mother nucleus is the only contributor to the initial state; thus, [7]

$$M_{f,i} = \int d^3r \left[\psi_D^* \phi_e^* \phi_\nu^* \right] \hat{V} \psi_M \tag{1.8}$$

While Fermi did not know the exact mathematical form of $\hat{V}$, through a consideration of the forms consistent with relativity, he determined that it would be a superposition of mathematical operators $\hat{O}_X$, where X gives the transformation properties and is either S (scalar), P (pseudoscalar), V (vector), A (axial vector), or T (tensor) [7].

The study of the symmetries and spatial properties of the decay products eventually led to the current understanding of the operator form, proposed by Feynman and Gell-Mann and Sudarshan and Marshak [9, 10]. Here, the operator has a structure of $\gamma_\mu(g_v \cdot 1 - g_A \gamma_5)$, only coupling to the vector and axial-vector interactions [11]. The strength of the coupling to each operator is measured by the coupling constants g_V and g_A. By the conserved vector current (CVC) hypothesis, $g_V = 1$ [12]. The partially conserved axial-vector current hypothesis gives $g_A = 1.25$ in free nucleons, though the many-nucleon correlations present in nuclei can reduce it by 20–30% [12]. The nuclear matrix elements can also be calculated for the individual operators; for the vector operator, this produces the Fermi matrix elements M_F, and for the axial-vector, the Gamow–Teller matrix elements M_{GT} [7].

The likeliness and consequently the decay rate of a β transition will depend on the orbital angular momentum of the emitted $e^{\pm}$ and ν_e or $\overline{\nu}_e$. In the most likely type of transition, an allowed β decay, the electron or positron and electron antineutrino or neutrino, which are intrinsically spin-1/2 particles, and are not emitted with any orbital angular momentum [7]. If the spins of the two emitted particles are anti-aligned, then the net spin is zero; this means that they carry away a total spin of $\vec{S} = \vec{0}$ from the decay, which is called a "Fermi decay" [12]. If instead the spins are parallel, the net spin is one, and they carry away a total spin of $\vec{S} = \vec{1}$; this is called a "Gamow–Teller" decay. For allowed transitions, a Fermi decay can thus only result in a change in the angular momentum of the nuclear states of $\Delta J = 0$, but a Gamow–Teller decay can result in a change of $\Delta J = 1$ or 0 [12]. Neither form of allowed decay can result in a change of parity π; furthermore, if the initial and final angular momentum of the nuclear state are zero, then only a Fermi decay is possible [7].

Thus, the criteria for an allowed decay is a change in angular momentum $\Delta J = 0$ or 1 and a change in parity $\Delta \pi = 0$. A decay that meets these criteria and also occurs between isobaric analog states, states with the same nuclear spin, parity, and total isospin T but different isospin projections T_z, is called a "superallowed" transitions [7]. These occur in two forms, those between $J^\pi = 0^+$ and 0^+ states, which must be purely Fermi, and thus these are called superallowed pure Fermi transitions, and those occurring between states of nonzero J, which can be either Fermi or Gamow–Teller, and are called superallowed mixed transitions.

1.2.2   Standard Model of the Electroweak Interaction

In the Standard Model, a description of electroweak decays requires the Cabibbo–Kobayashi–Maskawa (CKM) matrix, which was first proposed by Cabibbo to reconcile vector-current universality and decays involving strange quarks through a mixing between the first two generations of quarks [13]. The discovery of CP violation showed that this explanation was incomplete; Kobayashi and Maskawa demonstrated that the existence of a third generation of quarks and additional mixing in these decays provides an explanation [14]. The CKM matrix is a 3×3 unitary rotation matrix describing the mixing of the strong quark eigenstates under the weak interaction. By convention, the quarks of charge $+2/3$ (u, c, and t) are taken as unmixed, and the mixing is expressed as the CKM matrix acting on the quarks of charge $-1/3$ (d, s, and b); thus,

$$\begin{pmatrix} d' \\ s' \\ b' \end{pmatrix} = \begin{pmatrix} V_{ud} & V_{us} & V_{ub} \\ V_{cd} & V_{cs} & V_{cb} \\ V_{td} & V_{ts} & V_{tb} \end{pmatrix} \begin{pmatrix} d \\ s \\ b \end{pmatrix} \tag{1.9}$$

Here, d', s', and b' represent the weak eigenstates, superpositions of down-type, strange-type, and bottom-type quarks, respectively, and the various V_{ij} matrix elements represent the probability amplitudes of each type of decay between the strong (or mass) eigenstates [1, 2].

While this model does not provide a prediction for any of the individual values of the V_{ij} elements, it does require that the CKM matrix be unitary and to preserve norms across this transformation. If this is not the case, then either the theory of weak decays is incomplete—requiring the presence of other interactions such as scalar, pseudoscalar, or tensor interactions alongside the vector and axial-vector interactions—or it would indicate the presence of additional generation of quarks or other physics beyond the Standard Model [1]. While there are many ways to determine the unitarity of a matrix, the highest-precision result comes from the normalization of the top row of the CKM matrix [15]. This is done following:

$$|V_{ud}|^2 + |V_{us}|^2 + |V_{ub}|^2 = 1 \tag{1.10}$$

and is further simplified because the V_{ub} element is very small, meaning that only two elements contribute significantly to the precision of the overall normalization test [2]. These are V_{us}, which is calculated from kaon decays [4], and V_{ud}. This is an area of active study, and over the past few years, experiments have significantly improved the precision and accuracy of this normalization test [15–17]

1.2.3   Determining V_{ud}

The other significant contributor to the top-row normalization test is V_{ud}, which can be calculated following:

$$V_{ud} = \frac{G_V}{G_F} \tag{1.11}$$

where G_F is the weak-interaction constant for purely leptonic muon decays [2] and G_V is the vector coupling constant for semileptonic weak interactions [15]. As G_F is well-known [18, 19], the primary area of active research is the determination of V_{ud}. There are four primary methods used in this determination. Two of these, pion and neutron decays, offer relatively simple systems in which to observe V_{ud}.

Neutron decays have the advantage of determining G_V in a system free from any nuclear structure considerations. However, the accuracy of the value of G_V and thus V_{ud} from neutron decays is affected by technique-dependent conflicting measurements of the neutron lifetime [1] where different half-lives are determined from neutron beam [20] and trapped ultracold neutron [21–27] experiments. Pions also offer a nuclear-structure-free system for determining V_{ud}. The problem here lies in the very low branching ratio, on the order of 10^{-8}, of the pion β-decay, $\pi^+ \to \pi_0\, e^+\, \nu_e$. Thus, it also does not provide a particularly precise determination of V_{ud} [1].

Another possible way of determining V_{ud} comes from nuclear beta transitions. The conserved vector current (CVC) hypothesis asserts that the vector component of the semileptonic weak interaction, whose coupling constant is G_V, is unchanged by the presence of the strong force; this means that its measured value is independent of the nucleus in which it is measured. There are two beta decay systems from which G_V is currently determined, superallowed pure Fermi $0^+ \to 0^+$ transitions and superallowed mixed transitions in $T = 1/2$ isospin doublets in mirror nuclei [17, 28]. Of these, superallowed Fermi $0^+ \to 0^+$ transitions currently provide the most precise value for V_{ud} and thus the most stringent test of CKM unitarity. A comparison of the four methods for obtaining V_{ud} can be seen in Fig. 1.1.

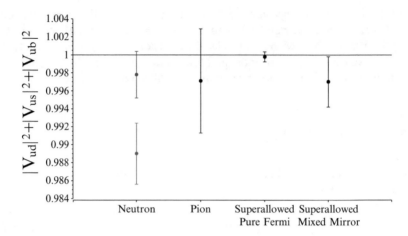

Fig. 1.1 Comparison of the top row normalization value for the CKM matrix calculated using the V_{ud} values from neutrons, recalculated [29] based on a recent measurement of the neutron β-decay asymmetry parameter [30] and either the beam [20] (red) or trapped ultracold neutron [21–27] (blue) half-lives; pion decays [1]; superallowed pure Fermi [17]; and superallowed mixed mirror [31–33] transitions

1.2.4 Superallowed Pure Fermi Transitions: $\mathcal{F}t^{0^+ \rightarrow 0^+}$

Superallowed Fermi $0^+ \rightarrow 0^+$ transitions are those that occur between isobaric analog states of spin and parity $J^\pi = 0^+$ and the same isospin T. The most precisely known transitions are those that occur between states of isospin $T = 1$. Since superallowed pure Fermi β decays only depend on the vector part of the weak interaction, the CVC hypothesis indicates that the ft-values, the statistical decay function, which is the product of the statistical rate function f and the partial half-life t, should be the same in all nuclei if we ignore nuclear interactions. Taking the definition of transition rate λ from Fermi's Golden Rule, Eq. (1.7), we can apply the definition of the transition rate in terms of the partial half-life t, $\lambda = \frac{\ln(2)}{t}$ and the definitions of the density of states to give us

$$\lambda = \frac{\ln(2)}{t} = \frac{2\pi}{\hbar} G_V^2 |M_F|^2 \frac{m_e^5 c^4}{4\pi^4 \hbar^6} f \tag{1.12}$$

This can then be rearranged to give [15]:

$$ft = \frac{K}{G_V^2 |M_F|^2} \tag{1.13}$$

Where $K/(\hbar c)^6 = 2\pi^3 \hbar \ln(2)/(m_e c^2)^5 = 8120.2776(9) \times 10^{-10}\,\mathrm{GeV^{-4}\,s}$ [17], G_V is the aforementioned semileptonic weak vector coupling constant, and M_F is the Fermi matrix element that for $T = 1$ decays is $M_F = \sqrt{2}$ [16]. This, however, ignores various radiative and structure-based corrections necessary to account for the nuclei in which these decays occur. Thus, the constant quantity is instead the corrected statistical decay function $\mathcal{F}t$ (or $\mathcal{F}t^{0^+ \to 0^+}$) [15]:

$$\mathcal{F}t^{0^+ \to 0^+} = ft \left(1 + \delta_R'\right) \left(1 + \delta_{NS}^V - \delta_C^V\right)$$

$$= \frac{K}{2G_V^2 \left(1 + \Delta_R^V\right)} = \frac{K}{2G_F^2 V_{ud}^2 \left(1 + \Delta_R^V\right)} \tag{1.14}$$

The various δ are theory-based corrections; δ_R', is a radiative correction dependent on the energy of the electron and the Z of the daughter nucleus [16, 31], obtained from QED calculations to each decay in question [17, 28]; δ_{NS}^V is another nuclear structure based correction [16, 31]; and δ_C^V is the isospin symmetry breaking correction [34]. Δ_R^V is the transition-independent radiative correction, for which the current best value is $\Delta_R^V = 2.361(38)\%$ [29].

There are three experimental quantities that go into the determination of V_{ud}: the Q value, the half-life $t_{1/2}$, and the branching ratio of the decay. The Q value is necessary for the calculation of the statistical rate function f which is an integral taken over phase space of the form [15]:

$$f = \int_1^{W_0} pW(W_0 - W)^2 F(Z, W) S(Z, W) dW \tag{1.15}$$

where W is the electron total energy in electron rest-mass units, $p = (W^2 - 1)^{1/2}$ is the electron momentum, Z is the atomic number of the daughter nucleus, $F(Z, W)$ is the Fermi function, and $S(Z, W)$ is the shape correction function. W_0 is the maximum value of W, which is calculable from the Q_{EC} value as $W_0 = \frac{Q_{EC}}{m_e} - 1$. To calculate f to a sufficient precision, many details of the motion of the decay must be considered. This includes electron wavefunctions that are the exact functions for the nuclear charge density distributions; lepton wavefunctions including second-forbidden corrections, relativistic corrections, and induced-current corrections (giving a nuclear structure dependence to the integral); and interactions with atomic electrons must be approximated with a screening correction [15]. Due to the difficulty of this calculation, Towner and Hardy, the compilers of the regular surveys of this field, have produced a parameterization of f for the decays of interest [35].

The half-life and branching ratio are used to calculate the partial half-life of the decay t, following:

$$t = \frac{t_{1/2}}{R}(1 + P_{EC}) \tag{1.16}$$

Where $t_{1/2}$ is the total half-life, R is the branching ratio, and P_{EC} is the electron capture fraction [15].

Historically, research has focused on the so-called traditional nine, those that decay to stable nuclei; these are ^{10}C, ^{14}O, ^{26m}Al, ^{34}Cl, ^{38m}K, ^{42}Sc, ^{46}V, ^{50}Mn, and ^{54}Co [15]. Continued data collection efforts have expanded the available candidates to fourteen, adding ^{22}Mg, ^{34}Ar, ^{38}Ca, ^{62}Ga, and ^{74}Rb [17]. The two lightest nuclei have been of recent interest, with a series of Q_{EC} value measurements on ^{10}C [36, 37] and ^{14}O [38], half-life measurements of ^{10}C [39] and ^{14}O [40], and branching ratio measurements on ^{14}O [41]. As illustrated in Fig. 1.1, the current most precise normalization test of the top row of the CKM matrix comes from these 14 measurements. Combined with the G_F value, this results in a V_{ud} value of $V_{ud} = 0.97412(21)$ and gives a unitarity test of the top row equal to 0.99978(55), which is consistent with 1 [17].

1.2.5 Superallowed Mixed Mirror Transitions: $\mathcal{F}t^{mirror}$

A complementary determination of V_{ud} from nuclear β decays is desirable to serve as a check on the value obtained from superallowed Fermi decays; to allow for testing of the methods used for calculating δ_c, which shows a large variation depending on the model used for the calculation [28, 34, 42]; and to test for unknown systematic effects or even new physics. One such method is the study of superallowed mixed mirror decays [28, 31]. Occurring between $T = 1/2$ isospin doublets in mirror nuclei, these transitions are mixed Fermi and Gamow–Teller decays, and thus have both vector and axial-vector contributions to the transition [31].

Because of the mixed nature of these transitions, the calculation of the corrected statistical decay function $\mathcal{F}t$ (here called $\mathcal{F}t^{mirror}$) is slightly different to the calculation presented in Eq. (1.14):

$$\mathcal{F}t^{mirror} = f_V t \left(1 + \delta_R'\right)\left(1 + \delta_{NS}^V - \delta_C^V\right) \tag{1.17}$$

Where the partial half-life t and the various corrections δ are the same, but f_v is the statistical rate function for only the vector part of this interaction [28]. Under the CVC hypothesis, this should remain the same for all the $T = 1/2$ superallowed mixed mirror decays.

The deconvolution of the vector and axial-vector components also results in a change to the right-hand side of Eq. (1.14). $\mathcal{F}t^{mirror}$ is related to the V_{ud} element of the CKM matrix by [28]:

$$\mathcal{F}t^{\text{mirror}} = \frac{K}{G_V^2} \frac{1}{|M_F^0|^2 C_V^2 \left(1 + \Delta_R^V\right)\left(1 + \frac{f_A}{f_V}\rho^2\right)}$$

$$= \frac{K}{G_F^2 V_{ud}^2} \frac{1}{|M_F^0|^2 C_V^2 \left(1 + \Delta_R^V\right)\left(1 + \frac{f_A}{f_V}\rho^2\right)}, \qquad (1.18)$$

where K, G_F, G_V, V_{ud}, and Δ_R are the same as before, M_F^0 is the Fermi matrix element in the isospin limit, which for these $T = 1/2$ mirror β decays is $|M_F^0|^2 = 1$, and $C_V^2 = 1$ is the vector coupling constant [28]. The quantity f_A is the statistical rate functions for the axial-vector parts of this interaction, and ρ is the Fermi-to-Gamow-Teller mixing ratio.

The experimental determination of ρ adds an additional experimental quantity necessary for the determination of V_{ud} to the three values necessary for superallowed $0^+ \rightarrow 0^+$ transitions. It can be determined from the measurement of either the β asymmetry parameter A_β, the β-neutrino angular correlation $a_{\beta\nu}$, or the neutrino asymmetry parameter B_ν. Currently, ρ has only been experimentally determined for five nuclei of interest, with ρ having been obtained from measurements of A_β for ^{19}Ne [43], ^{29}P [44], and ^{35}Ar [45, 46]; from measurements of B_ν for ^{37}K [33, 47]; and from measurements of $a_{\beta\nu}$ for ^{21}Na [48]. This limits the ensemble of transitions which can be considered for the determination of V_{ud}, though efforts are underway to expand this list, including measuring A_β in ^{23}Mg using versatile ion-polarized techniques online (VITO) at ISOLDE [49] and a new ion trapping experiment under development at the Nuclear Science Laboratory (NSL) at the University of Notre Dame to measure $a_{\beta\nu}$ for lighter superallowed mixed mirror transitions [50].

Ongoing efforts to improve the other experimental data that go into determining V_{ud} via superallowed mixed mirrors continue. Recent efforts have included measuring Q_{EC} values using Penning trap mass spectrometry for increased precision; measurements have occurred on ^{21}Na [32], ^{29}P [32], and ^{11}C [51]. New half-life measurements have occurred on ^{17}F [52, 53], ^{19}Ne [54, 55], ^{21}Na [52], ^{25}Al [56], ^{33}Cl [57], and ^{37}K [58], and a higher-precision value of ρ for ^{37}K has been determined through a measurement of A_β [33]. Chapter 2 of this thesis details the new high-precision half-life measurement of ^{11}C. The contributions to the overall uncertainty on the $\mathcal{F}t^{\text{mirror}}$ value for each experimental and theoretical parameter can be seen in Fig. 1.2, where the isotopes for which experimental parameters remain the limiting factor on the precision can be seen. From the five decays for which all the necessary information is known, the current data give a value of $V_{ud} = 0.9727(14)$ and a unitarity test of the top row equal to $0.9970(28)$ [31–33]. These values are just over 1σ away from the values given by superallowed $0^+ \rightarrow 0^+$ decays and unity. In order to further accentuate possible evidence of physics beyond the Standard Model, more measurements and greater precision are needed.

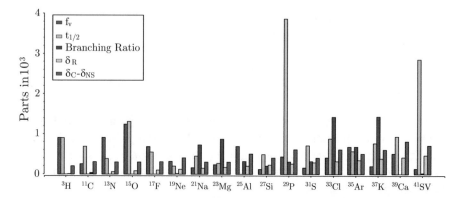

Fig. 1.2 Figure showing the relative uncertainty in the three experimental parameters going into the determination of $\mathcal{F}t^{\mathrm{mirror}}$ (the statistical rate function f_V (calculated from Q_{EC} values), the half-life $t_{1/2}$, and the branching ratio) and the two combined theoretical corrections (δ'_R, the radiative correction, and $\delta^V_{NS} - \delta^V_C$, the difference between the isospin symmetry breaking correction and the nuclear structure based correction)

1.3   Determining the *rp*-Process Path with Penning Trap Mass Spectrometry

1.3.1   *rp-Process*

Type I X-ray bursts are astronomical events that occur in binary systems where a neutron star accretes hydrogen and helium-rich material from its companion star; the accretion of more matter on the surface of the neutron star results in increasing densities and temperatures until the accreted material undergoes thermonuclear runaway [59]. The energy generated during this process gives rise to an increase in temperature and sharp increase of X-ray luminosity followed by a slower decay as the atmosphere cools.

The high temperatures and densities achieved during this event provide the conditions necessary to trigger the rapid proton capture (*rp*) process, a nuclear burning process for proton-rich nuclei lighter than $A \sim 106$ [60, 61]. The *rp*-process is dominated by a sequence of rapid proton captures and β decays along the proton dripline. It begins with a breakout from the hot CNO cycle through α capture reactions [60], and then branches away from (α,p)-process at one or several points via proton capture reactions [62].

The *rp*-process flows through a series of proton capture (p,γ), photodisintegration (γ,p), α capture (α,p), and β^+-decay reactions, with relative rates of reactions determining the pathway. Type I X-ray bursts generally have rise times of $\sim$1–10 s, and decay times ranging from 10 s to several minutes, though much longer-lived superbursts, with hour-long decay times, also exist [63]. Of particular importance in determining the *rp*-process flow is the ratio of the (p,γ) and (γ,p) reaction rates, which are highly sensitive to the Q values of these reactions [64].

1.3.2   Reaction Rate

In an isotonic or isotopic equilibrium, the abundance ratio of two neighboring nuclei n and $n + 1$ is given by the Saha equation [65, 66]:

$$\frac{Y_{n+1}}{Y_n} = \rho_n \frac{G_{n+1}}{G_n} \left(\frac{A_{n+1}}{A_n} \frac{2\pi\hbar^2}{m_u kT} \right)^{3/2} \exp\left(\frac{Q}{kT} \right) \tag{1.19}$$

Where Y_n and Y_{n+1} are the abundances of the two neighboring nuclei, ρ_n is the proton or neutron density, G are the partition functions, A the mass numbers, m_u is the atomic mass unit, k the Boltzmann constant, T is the temperature, and Q is the relevant Q-value of the reaction, either the proton or neutron separation energy. Since the proton or neutron separation energies are calculated from the masses of the two nuclei and the proton or neutron captured, this clearly demonstrates an exponential dependence on mass. However, the *rp*-process does not reach (p,γ)–(γ,p) equilibrium for all reactions, nor does it occur across the whole of a Type I X-ray burst for all reactions. For example, in the ^{64}Ge(p, γ)^{65}As(p, γ)^{66}Se reaction pathway, the small proton separation energy of ^{65}As means the ^{65}As(γ,p) reaction is fast and thus (p,γ)–(γ,p) equilibrium is established throughout the reaction flow through this pathway, while the large proton separation energy ^{66}Se means that temperatures in excess of 1.5 GK are necessary to establish (p,γ)–(γ,p) equilibrium [61].

Thus, to determine the reaction flow of the *rp*-process in an X-ray burst, the reaction rates must be calculated. Resonant proton capture rates, which describes most of the relevant nuclear reactions in the *rp*-process, can be approximated by [66]:

$$N_A \langle \sigma v \rangle \propto \sum_i (\omega\gamma)_i \exp\left(-E_i/kT \right) \tag{1.20}$$

where $E_i = E_i^x - Q$ is the ith resonance for excitation energy E_i^x, Q is the Q value of the reaction, the difference in mass between the initial and final states, and $(\omega\gamma)_i$ is the ith resonance strength, determined by:

$$(\omega\gamma)_i = \frac{2J_i + 1}{(2J_p + 1)(2J_T + 1)} \frac{\Gamma_p \Gamma_\gamma}{\Gamma_p + \Gamma_\gamma} \tag{1.21}$$

where J_i, J_p, and J_T are the spins of the resonance, proton, and ground-state proton-capturing nucleus, respectively, and Γ_γ and Γ_p are the γ and proton partial widths.

This rate calculation can also be seen to be exponentially dependent on the Q-value of the reaction and thus of the mass.

1.3.3 rp-Process Waiting Points

From the exponential relationship of the (p, γ) reaction rate to the Q value of the reaction, it can be seen that a low proton-capture Q-value would result in a reduced reaction rate. Bottlenecks in the *rp*-process occur where low proton-capture Q values make the forward and reverse reaction rates competitive and β^+ decays or electron capture become the dominant route. Where this half-life is long, relative to the timescale of the X-ray burst, a waiting point occurs.

Accurately determining the reaction rates in and near these waiting points is particularly critical for determining the reaction flow of the *rp*-process, which requires the measurement of the experimental quantities involved in calculating the reaction rates, most significantly the masses [61]. The accurate calculations of light curves and isotopic abundances in the ashes of X-ray bursts also rely on the accurate understanding of the reaction pathway taken by the *rp*-process [67]; thus, precision mass measurements of these isotopes are necessary. Chapters 3 and 4 discuss Penning trap mass spectrometry and its use to measure the mass of ^{56}Cu, of interest for determining the reaction flow around the ^{56}Ni waiting point.

1.4 A New Facility for Precision *r*-Process Measurements

1.4.1 r-Process

The rapid neutron capture process or *r*-process is believed to account for approximately half of all nuclei heavier than the iron peak [68]. It proceeds through a series of rapid neutron captures away from stability, followed by beta decays back towards stability, illustrated in Fig. 1.3. The precise astrophysical site remains a source of contention; the *r*-process requires high temperatures and neutron fluxes, with various promising sites like the neutrino-driven winds of core-collapse supernovae [68, 69], the magneto-hydrodynamic jets of rotating supernovae [70], or neutron star mergers [71] having been proposed. The recent multi-messenger observation of a neutron star merger [72] through gravitational wave signal GW170817 [73] and accompanying kilonova AT2017gfo [74, 75] has however provided direct evidence of *r*-process nucleosynthesis [76, 77].

Indeed, comparison of *r*-process abundances from models and observation will play a critical role in determining the location of the *r*-process. Recent *r*-process sensitivity studies have shown that among the quantities that go into the calculation of reaction rate calculations—which includes neutron-capture cross-

Fig. 1.3 Illustration of the
reaction flow of the r-process
along the chart of the nuclides

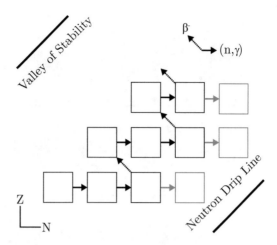

sections, beta-decay half-lives, beta-delayed neutron emission branching rates, and fission probabilities—the mass-derived neutron separation energies are those to which the final abundances are most sensitive [78]. However, because the r-process path is far from the valley of stability, the direct measurement of these masses is currently impossible, and so instead theoretical mass models such as the finite-range droplet model (FRDM) [79], the Weisäcler-Skyrme (WS) model [80, 81], the KTUY05 [82] and Duflo-Zuker (DZ) [83, 84] empirical formulae, and mass formulae based on the Hartree–Fock–Bogoliubov (HFB) approach [85] need to be used. As Fig. 1.4[1] (from [78]) shows, these formulae (FRDM1995, FRDM2012 [86], WS3, KTUY05, DZ33, HFB-17 [87], and HFB-24 [88] shown) all generally agree where mass data currently exists; however, as seen in Fig. 1.5,[2] there is considerable variation as they get farther from the known masses in the atomic mass evaluation (here showing AME2012 [89]).

1.4.2 Mass Sensitivity Studies

A series of simulations using a complete, dynamical r-process model have recently been performed to determine which mass uncertainties have the greatest influence on the r-process abundance [78]. This showed that well-known r-process mass abundance peaks at $A \approx 130$ and $A \approx 195$ are strongly influenced by masses near the appropriately closed $N = 82$ and $N = 126$ closed neutron shells, as well as identifying masses of importance for the rare-earth peak ($A \approx 165$); the relative dearth of experimental data for the later two regions has a particularly strong impact,

---

[1] Reprinted from Mumpower et al. [78], Copyright 2016, with permission from Elsevier.
[2] Reprinted from Mumpower et al. [78], Copyright 2016, with permission from Elsevier.

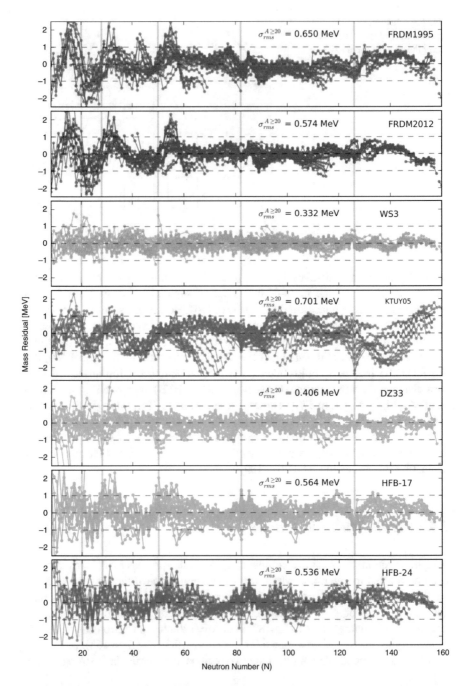

Fig. 1.4 Figure 5 from [78] comparing predictions of seven commonly used mass models (FRDM1995 [79], FRDM2012 [86], WS3 [80, 81], KTUY05 [82], DZ33 [83, 84], HFB-17 [87] and HFB-24 [88]) with the experimental values from the 2012 atomic mass evaluation (AME2012) [89]

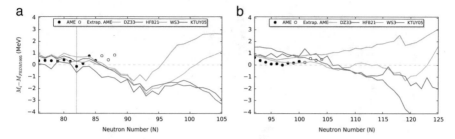

Fig. 1.5 Figure 6 from [78] showing comparisons of measured and extrapolated masses from AME2012 [89] and the predicted masses from DZ33 [83, 84], HFB21 [88], WS3 [80, 81], and KTUY05 [82] to those predicted by FRDM1995 [79] for (**a**) Tin ($Z = 50$) and (**b**) Europium ($Z = 63$)

as illustrated in Fig. 1.6,[3] where darker colors indicate a stronger impact on the final calculated abundances.

Recently, mass measurements have focused on filling in these missing data. Penning trap mass measurements of neutron-rich cadmium using ISOLTRAP at ISOLDE–CERN [90] and of neutron-rich indium using the TITAN Penning trap at TRIUMF [91] have provided considerable data around the $N = 82$ shell closure. Ongoing Penning trap mass spectrometry research campaigns are focused on filling in the unknown masses for the rare-earth peak. These efforts use the CARIBU (Californium Rare Isotope Breeder Upgrade) facility, based on ^{252}Cf spontaneous fission, and the Canadian Penning Trap (CPT) at Argonne National Laboratory [92, 93] and proton-induced Uranium fission at the University of Jyväskylä's IGISOL facility using JYFLTRAP [94].

1.4.3 Need for a New Facility

There is currently no equivalent ongoing effort to measure masses at the $N = 126$ neutron shell closure. This is because it is beyond the reach of current accelerator facilities, which primarily use projectile-fragmentation, target-fragmentation, or fission to produce rare isotope beam. All of these either lack the relevant beams or targets or have production cross-sections of the isotopes of interest in the $N = 126$ region that are too low to allow for mass measurements [95]. Thus, an alternate production mechanism is needed.

Such a mechanism, multi-nuclear transfer reactions (MNTs), was proposed by Dasso, Pollarolo, and Winther for future accelerator facilities [96] and by Zagrebaev and Greiner for production in the $N = 126$ region [97]. It relies on the transfer of multiple nucleons between heavy beams and heavy targets in deep inelastic

---

[3]Reprinted from Mumpower et al. [78], Copyright 2016, with permission from Elsevier.

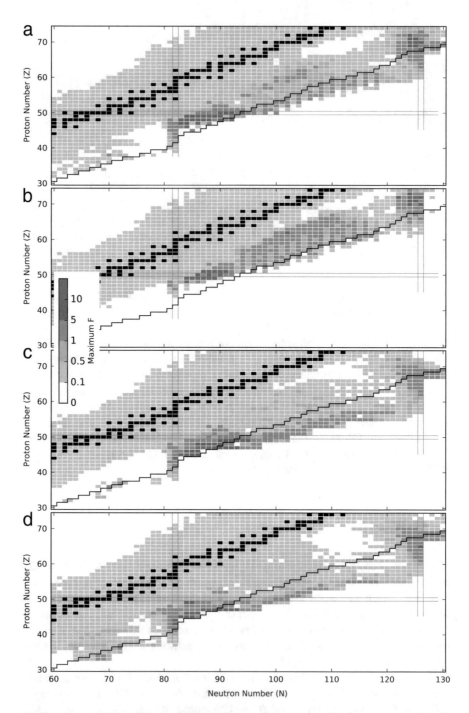

Fig. 1.6 Figure 13 from [78], showing nuclei that significantly impact final *r*-process abundances under four different astrophysical conditions, (**a**) low entropy hot wind, (**b**) high entropy hot wind, (**c**) cold wind, and (**d**) neutron star merger. More influential nuclei are shaded darker based on the impact parameter *F* from ±500 keV mass variation of the nuclei. Light gray shading indicates the extent of measured masses from the AME2012 [89], and the black line indicates an estimate of neutron-rich availability from FRIB

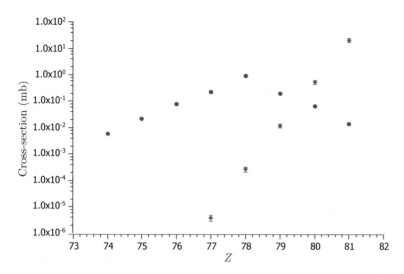

Fig. 1.7 Comparison of the measured cross-sections along $N = 126$ from the MNT reaction ^{136}Xe$+^{198}$Pt at GANIL [95] (blue circles) and the projectile fragmentation reaction of ^{208}Pb$+^{9}$Be at GSI [100] (red diamonds)

collisions near the Coulomb barrier. The production of isotopes of interest by such reactions was demonstrated by recent efforts using the EXOGAM high-efficiency germanium array at GANIL for use in the KEK isotope separator at RIKEN [98, 99]. As seen in Fig. 1.7, the ^{136}Xe beam—^{198}Pt target reaction [95] has a significantly higher production cross-section than projectile fragmentation using ^{208}Pb and ^{9}Be [100] for $N = 126$ isotopes of interest; the GRAZING code [101, 102] also calculates a larger cross-section for MNT reactions than the measured fragmentation reaction [103]. A new facility, the $N = 126$ factory, is under development at the Argonne Tandem Linac Accelerator System (ATLAS) that will make use of these reactions to study masses in the $N = 126$ region. Chapter 5 will detail the design of this facility and specifically discuss the commissioning of a radiofrequency quadrupole (RFQ) cooler-buncher, a critical component that will provide the required bunched ions for mass measurement using the CPT.

References

1. N. Severijns, O. Naviliat-Cuncic, Symmetry tests in nuclear beta decay. Annu. Rev. Nucl. Part. S. **61**, 23–46 (2011)
2. J.C. Hardy, I.S. Towner, CKM unitarity normalization tests, present and future. Ann. Phys. (Berl.) **525**, 443–451 (2013)
3. J.D. Hobbs, M.S. Neubauer, S. Willenbrock, Tests of the standard electroweak model at the energy frontier. Rev. Mod. Phys. **84**, 1477–1526 (2012)
4. O. Naviliat-Cuncic, M. González-Alonso, Prospects for precision measurements in nuclear β decay in the LHC era. Ann. Phys. (Berl.) **525**, 600–619 (2013)

5. J. Hewett, H. Weerts, R. Brock, J. Butler, B. Casey, J. Collar, A. de Gouvea, R. Essig, Y. Grossman, W. Haxton et al., Fundamental physics at the intensity frontier, arXiv preprint arXiv:1205.2671 (2012)
6. K. Blaum, High-accuracy mass spectrometry with stored ions. Phys. Rep. **425**, 1–78 (2006)
7. K.S. Krane, D. Halliday, *Introductory Nuclear Physics* (Wiley, New York, 1988)
8. E. Fermi, An attempt of a theory of beta radiation. 1. Z. Phys. **88**, 10 (1934)
9. R.P. Feynman, M. Gell-Mann, Theory of the Fermi interaction. Phys. Rev. **109**, 193–198 (1958)
10. E.C.G. Sudarshan, R.E. Marshak, Chirality invariance and the universal Fermi interaction. Phys. Rev. **109**, 1860–1862 (1958)
11. E.D. Commins, P.H. Bucksbaum, *Weak Interactions of Leptons and Quarks* (Cambridge University Press, Cambridge, 1983)
12. J. Suhonen, *From Nucleons to Nucleus: Concepts of Microscopic Nuclear Theory* (Springer Science & Business Media, Berlin, 2007)
13. N. Cabibbo, Unitary symmetry and leptonic decays. Phys. Rev. Lett. **10**, 531–533 (1963)
14. M. Kobayashi, T. Maskawa, CP-violation in the renormalizable theory of weak interaction. Progr. Theor. Phys. **49**, 652–657 (1973)
15. J.C. Hardy, I.S. Towner, Superallowed $0^+ \rightarrow 0^+$ nuclear β decays: a critical survey with tests of the conserved vector current hypothesis and the standard model. Phys. Rev. C **71**, 055501 (2005)
16. J.C. Hardy, I.S. Towner, Superallowed $0^+ \rightarrow 0^+$ nuclear β decays: a new survey with precision tests of the conserved vector current hypothesis and the standard model. Phys. Rev. C **79**, 055502 (2009)
17. J.C. Hardy, I.S. Towner, Superallowed $0^+ \rightarrow 0^+$ nuclear β decays: 2014 critical survey, with precise results for V_{ud} and CKM unitarity. Phys. Rev. C **91**, 025501 (2015)
18. V. Tishchenko, S. Battu, R.M. Carey, D.B. Chitwood, J. Crnkovic, P.T. Debevec, S. Dhamija, W. Earle, A. Gafarov, K. Giovanetti, T.P. Gorringe, F.E. Gray, Z. Hartwig, D.W. Hertzog, B. Johnson, P. Kammel, B. Kiburg, S. Kizilgul, J. Kunkle, B. Lauss, I. Logashenko, K.R. Lynch, R. McNabb, J.P. Miller, F. Mulhauser, C.J.G. Onderwater, Q. Peng, J. Phillips, S. Rath, B.L. Roberts, D.M. Webber, P. Winter, B. Wolfe, Detailed report of the MuLan measurement of the positive muon lifetime and determination of the Fermi constant. Phys. Rev. D **87**, 052003 (2013)
19. K. Olive, Particle Data Group, Review of particle physics. Chin. Phys. C **38**, 090001 (2014)
20. A.T. Yue, M.S. Dewey, D.M. Gilliam, G.L. Greene, A.B. Laptev, J.S. Nico, W.M. Snow, F.E. Wietfeldt, Improved determination of the neutron lifetime. Phys. Rev. Lett. **111**, 222501 (2013)
21. A.P. Serebrov, E.A. Kolomensky, A.K. Fomin, I.A. Krasnoshchekova, A.V. Vassiljev, D.M. Prudnikov, I.V. Shoka, A.V. Chechkin, M.E. Chaikovskiy, V.E. Varlamov, S.N. Ivanov, A.N. Pirozhkov, P. Geltenbort, O. Zimmer, T. Jenke, M. Van der Grinten, M. Tucker, Neutron lifetime measurements with a large gravitational trap for ultracold neutrons. Phys. Rev. C **97**, 055503 (2018)
22. R.W. Pattie, N.B. Callahan, C. Cude-Woods, E.R. Adamek, L.J. Broussard, S.M. Clayton, S.A. Currie, E.B. Dees, X. Ding, E.M. Engel, D.E. Fellers, W. Fox, P. Geltenbort, K.P. Hickerson, M.A. Hoffbauer, A.T. Holley, A. Komives, C.-Y. Liu, S.W.T. MacDonald, M. Makela, C.L. Morris, J.D. Ortiz, J. Ramsey, D.J. Salvat, A. Saunders, S.J. Seestrom, E.I. Sharapov, S.K. Sjue, Z. Tang, J. Vanderwerp, B. Vogelaar, P.L. Walstrom, Z. Wang, W. Wei, H.L. Weaver, J.W. Wexler, T.L. Womack, A.R. Young, B.A. Zeck, Measurement of the neutron lifetime using a magneto-gravitational trap and in situ detection. Science **360**, 627–632 (2018)
23. S. Arzumanov, L. Bondarenko, S. Chernyavsky, P. Geltenbort, V. Morozov, V. Nesvizhevsky, Y. Panin, A. Strepetov, A measurement of the neutron lifetime using the method of storage of ultracold neutrons and detection of inelastically up-scattered neutrons. Phys. Lett. B **745**, 79–89 (2015)

24. A. Steyerl, J.M. Pendlebury, C. Kaufman, S.S. Malik, A.M. Desai, Quasielastic scattering in the interaction of ultracold neutrons with a liquid wall and application in a reanalysis of the Mambo I neutron-lifetime experiment. Phys. Rev. C **85**, 065503 (2012)
25. A. Pichlmaier, V. Varlamov, K. Schreckenbach, P. Geltenbort, Neutron lifetime measurement with the UCN trap-in-trap MAMBO II. Phys. Lett. B **693**, 221–226 (2010)
26. A. Serebrov, V. Varlamov, A. Kharitonov, A. Fomin, Y. Pokotilovski, P. Geltenbort, J. Butterworth, I. Krasnoschekova, M. Lasakov, R. Tal'daev, A. Vassiljev, O. Zherebtsov, Measurement of the neutron lifetime using a gravitational trap and a low-temperature Fomblin coating. Phys. Lett. B **605**, 72–78 (2005)
27. W. Mampe, L. Bondarenko, V. Morozov, Y.N. Panin, A. Fomin, Measuring neutron lifetime by storing ultracold neutrons and detecting inelastically scattered neutrons. JETP Lett. **57**, 82–82 (1993)
28. N. Severijns, M. Tandecki, T. Phalet, I.S. Towner, $\mathcal{F}t$ values of the $T = 1/2$ mirror β transitions. Phys. Rev. C **78**, 055501 (2008)
29. W.J. Marciano, A. Sirlin, Improved calculation of electroweak radiative corrections and the value of V_{ud}. Phys. Rev. Lett. **96**, 032002 (2006)
30. M.A.-P. Brown, E.B. Dees, E. Adamek, B. Allgeier, M. Blatnik, T.J. Bowles, L.J. Broussard, R. Carr, S. Clayton, C. Cude-Woods, S. Currie, X. Ding, B.W. Filippone, A. García, P. Geltenbort, S. Hasan, K.P. Hickerson, J. Hoagland, R. Hong, G.E. Hogan, A.T. Holley, T.M. Ito, A. Knecht, C.-Y. Liu, J. Liu, M. Makela, J.W. Martin, D. Melconian, M.P. Mendenhall, S.D. Moore, C.L. Morris, N. Nepal, N. Nouri, R.W. Pattie, A. Pérez Galván, D.G. Phillips, R. Picker, M.L. Pitt, B. Plaster, J.C. Ramsey, R. Rios, D.J. Salvat, A. Saunders, W. Sondheim, S.J. Seestrom, S. Sjue, S. Slutsky, X. Sun, C. Swank, G. Swift, E. Tatar, R.B. Vogelaar, B. VornDick, Z. Wang, J. Wexler, T. Womack, C. Wrede, A.R. Young, B.A. Zeck, New result for the neutron β-asymmetry parameter A_0 from UCNA. Phys. Rev. C **97**, 035505 (2018)
31. O. Naviliat-Cuncic, N. Severijns, Test of the conserved vector current hypothesis in $T = 1/2$ mirror transitions and new determination of $|V_{ud}|$. Phys. Rev. Lett. **102**, 142302 (2009)
32. M. Eibach, G. Bollen, M. Brodeur, K. Cooper, K. Gulyuz, C. Izzo, D.J. Mor-rissey, M. Redshaw, R. Ringle, R. Sandler, S. Schwarz, C.S. Sumithrarachchi, A.A. Valverde, A.C.C. Villari, Determination of the Q_{EC} values of the $T = 1/2$ mirror nuclei ^{21}Na and ^{29}P at LEBIT. Phys. Rev. C **92**, 045502 (2015)
33. B. Fenker, A. Gorelov, D. Melconian, J.A. Behr, M. Anholm, D. Ashery, R.S. Behling, I. Cohen, I. Craiciu, G. Gwinner, J. McNeil, M. Mehlman, K. Olchanski, P.D. Shidling, S. Smale, C.L. Warner, Precision measurement of the β asymmetry in spin-polarized ^{37}K decay. Phys. Rev. Lett. **120**, 062502 (2018)
34. I.S. Towner, J.C. Hardy, Improved calculation of the isospin-symmetry-breaking corrections to superallowed Fermi β decay. Phys. Rev. C **77**, 025501 (2008)
35. I.S. Towner, J.C. Hardy, Parametrization of the statistical rate function for select superallowed transitions. Phys. Rev. C **91**, 015501 (2015)
36. T. Eronen, D. Gorelov, J. Hakala, J.C. Hardy, A. Jokinen, A. Kankainen, V.S. Kolhinen, I.D. Moore, H. Penttilä, M. Reponen, J. Rissanen, A. Saastamoinen, J. Äystö, Q_{EC} values of the superallowed β emitters ^{10}C, ^{34}Ar, ^{38}Ca, and ^{46}V. Phys. Rev. C **83**, 055501 (2011)
37. A.A. Kwiatkowski, A. Chaudhuri, U. Chowdhury, A.T. Gallant, T.D. Macdonald, B.E. Schultz, M.C. Simon, J. Dilling, Mass measurements of singly and highly charged radioactive ions at TITAN: a new Q_{EC}-value measurement of ^{10}C. Ann. Phys. (Berl.) **525**, 529–537 (2013)
38. A.A. Valverde, G. Bollen, M. Brodeur, R.A. Bryce, K. Cooper, M. Eibach, K. Gulyuz, C. Izzo, D.J. Morrissey, M. Redshaw, R. Ringle, R. Sandler, S. Schwarz, C.S. Sumithrarachchi, A.C.C. Villari, First direct determination of the superallowed β-decay Q_{EC} value for ^{14}O. Phys. Rev. Lett. **114**, 232502 (2015)

39. M.R. Dunlop, C.E. Svensson, G.C. Ball, G.F. Grinyer, J.R. Leslie, C. Andreoiu, R.A.E. Austin, T. Ballast, P.C. Bender, V. Bildstein, A. Diaz Varela, R. Dunlop, A.B. Garnsworthy, P.E. Garrett, G. Hackman, B. Hadinia, D.S. Jamieson, A.T. Laffoley, A.D. MacLean, D.M. Miller, W.J. Mills, J. Park, A.J. Radich, M.M. Rajabali, E.T. Rand, C. Unsworth, A. Valencik, Z.M. Wang, E.F. Zganjar, High-precision half-life measurements for the superallowed β^+ emitter ^{10}C: implications for weak scalar currents. Phys. Rev. Lett. **116**, 172501 (2016)

40. A.T. Laffoley, C.E. Svensson, C. Andreoiu, R.A.E. Austin, G.C. Ball, B. Blank, H. Bouzomita, D.S. Cross, A. Diaz Varela, R. Dunlop, P. Finlay, B. Garnsworthy, P.E. Garrett, J. Giovinazzo, G.F. Grinyer, G. Hackman, B. Hadinia, D.S. Jamieson, S. Ketelhut, K.G. Leach, J.R. Leslie, E. Tardiff, J.C. Thomas, C. Unsworth, High-precision half-life measurements for the superallowed Fermi β^+ emitter ^{14}O. Phys. Rev. C **88**, 015501 (2013)

41. P.A. Voytas, E.A. George, G.W. Severin, L. Zhan, L.D. Knutson, Measurement of the branching ratio for the β decay of ^{14}O. Phys. Rev. C **92**, 065502 (2015)

42. W. Satuła, J. Dobaczewski, W. Nazarewicz, T.R. Werner, Isospin-breaking corrections to superallowed Fermi β decay in isospin- and angular-momentum-projected nuclear density functional theory. Phys. Rev. C **86**, 054316 (2012)

43. F.P. Calaprice, S.J. Freedman, W.C. Mead, H.C. Vantine, Experimental study of weak magnetism and second-class interaction effects in the β decay of polarized ^{19}Ne. Phys. Rev. Lett. **35**, 1566–1570 (1975)

44. G.S. Masson, P.A. Quin, Measurement of the asymmetry parameter for ^{29}P β decay. Phys. Rev. C **42**, 1110–1119 (1990)

45. J.D. Garnett, E.D. Commins, K.T. Lesko, E.B. Norman, β-Decay asymmetry parameter for ^{35}Ar: an anomaly resolved. Phys. Rev. Lett. **60**, 499–502 (1988)

46. A. Converse, M. Allet, W. Haeberli, W. Hajdas, J. Lang, J. Liechti, H. Lüscher, M. Miller, R. Müller, O. Naviliat-Cuncic, P. Quin, J. Sromicki, Measurement of the asymmetry parameter for ^{35}ar β-decay as a test of the CVC hypothesis. Phys. Lett. B **304**, 60–64 (1993)

47. D. Melconian, J. Behr, D. Ashery, O. Aviv, P Bricault, M. Dombsky S. Fostner, A. Gorelov, S. Gu, V. Hanemaayer, K. Jackson, M. Pearson, I. Vollrath, Measurement of the neutrino asymmetry in the β decay of laser-cooled, polarized ^{37}K. Phys. Lett. B **649**, 370–375 (2007)

48. P.A. Vetter, J.R. Abo-Shaeer, S.J. Freedman, R. Maruyama, Measurement of the β-ν correlation of ^{21}Na using shakeoff electrons. Phys. Rev. C **77**, 035502 (2008)

49. M. Deicher, M. Stachura, V. Amaral, M. Bissell, J. Correia, A. Gottberg, L. Hemmingsen, S. Hong, K. Johnston, Y. Kadi, M. Kowalska, J. Lehnert, A. Lopes, G. Neyens, K. Potzger, D. Pribat, N. Severijns, C. Tenreiro, P. Thulstrup, T. Trindade, T. Wichert, H. Wolf, D. Yordanov, Z. Salman, VITO – Versatile Ion-Polarized Techniques On-Line at ISOLDE (Former ASPIC UHV Beamline), Technical Report CERN-INTC-2013-013. INTC-O-017 (CERN, Geneva, 2013)

50. M. Brodeur, J. Kelly, J. Long, C. Nicoloff, B. Schultz, V_{ud} determination from light nuclide mirror transitions. Nucl. Instrum. Meth. Phys. Res. B **376**; Proceedings of the XVIIth International Conference on Electromagnetic Isotope Separators and Related Topics (EMIS2015), 11–15 May 2015 (Grand Rapids, MI, 2016), pp. 281–283

51. K. Gulyuz, G. Bollen, M. Brodeur, R.A. Bryce, K. Cooper, M. Eibach, C. Izzo, E. Kwan, K. Manukyan, D.J. Morrissey, O. Naviliat-Cuncic, M. Redshaw, R. Ringle, R. Sandler, S. Schwarz, C.S. Sumithrarachchi, A.A. Valverde, A.C.C. Villari, High precision determination of the β decay Q_{EC} value of ^{11}C and implications on the tests of the standard model. Phys. Rev. Lett. **116**, 012501 (2016)

52. J. Grinyer, G.F. Grinyer, M. Babo, H. Bouzomita, P. Chauveau, P. Delahaye, M. Dubois, R. Frigot, P. Jardin, C. Leboucher, L. Maunoury, C. Seiffert, J.C. Thomas, E. Traykov, High-precision half-life measurement for the isospin $T = 1/2$ mirror β^+ decay of ^{21}Na. Phys. Rev. C **91**, 032501 (2015)

53. M. Brodeur, C. Nicoloff, T. Ahn, J. Allen, D.W. Bardayan, F.D. Becchetti, Y.K. Gupta, M.R. Hall, O. Hall, J. Hu, J.M. Kelly, J.J. Kolata, J. Long, P. O'Malley, B.E. Schultz, Precision half-life measurement of ^{17}F. Phys. Rev. C **93**, 025503 (2016)

54. S. Triambak, P. Finlay, C.S. Sumithrarachchi, G. Hackman, G.C. Ball, P.E. Garrett, C.E. Svensson, D.S. Cross, A.B. Garnsworthy, R. Kshetri, J.N. Orce, M.R. Pearson, E.R. Tardiff, H. Al-Falou, R.A.E. Austin, R. Churchman, M.K. Djongolov, R. D'Entremont, C. Kierans, L. Milovanovic, S. O'Hagan, S. Reeve, S.K.L. Sjue, S.J. Williams, High-precision measurement of the ^{19}Ne half-life and implications for right-handed weak currents. Phys. Rev. Lett. **109**, 042301 (2012)

55. L.J. Broussard, H.O. Back, M.S. Boswell, A.S. Crowell, P. Dendooven, G.S. Giri, C.R. Howell, M.F. Kidd, K. Jungmann, W.L. Kruithof, A. Mol, C.J.G. Onderwater, R.W. Pattie, P.D. Shidling, M. Sohani, D.J. van der Hoek, A. Rogachevskiy, E. Traykov, O.O. Versolato, L. Willmann, H.W. Wilschut, A.R. Young, Measurement of the half-life of the $T = \frac{1}{2}$ mirror decay of ^{19}Ne and its implication on physics beyond the standard model. Phys. Rev. Lett. **112**, 212301 (2014)

56. J. Long, T. Ahn, J. Allen, D.W. Bardayan, F.D. Becchetti, D. Blankstein, M. Brodeur, D. Burdette, B. Frentz, M.R. Hall, J.M. Kelly, J.J. Kolata, P.D. O'Malley, B.E. Schultz, S.Y. Strauss, A.A. Valverde, Precision half-life measurement of ^{25}Al. Phys. Rev. C **96**, 015502 (2017)

57. J. Grinyer, G.F. Grinyer, M. Babo, H. Bouzomita, P. Chauveau, P. Delahaye, M. Dubois, R. Frigot, P. Jardin, C. Leboucher, L. Maunoury, C. Seiffert, J.C. Thomas, E. Traykov, High-precision half-life measurements of the $T = 1/2$ mirror β decays ^{17}F and ^{33}Cl. Phys. Rev. C **92**, 045503 (2015)

58. P.D. Shidling, D. Melconian, S. Behling, B. Fenker, J.C. Hardy, V.E. Iacob, E. McCleskey, M. McCleskey, M. Mehlman, H.I. Park, B.T. Roeder, Precision half-life measurement of the β^+ decay of ^{37}K. Phys. Rev. C **90**, 032501 (2014)

59. S.E. Woosley, R.E. Taam, Gamma-ray bursts from thermonuclear explosions on neutron stars. Nature **263**, 101–103 (1976)

60. R.K. Wallace, S.E. Woosley, Explosive hydrogen burning. Astrophys. J. Suppl. Ser. **45**, 389–420 (1981)

61. H. Schatz, A. Aprahamian, V. Barnard, L. Bildsten, A. Cumming, M. Ouel-lette, T. Rauscher, F.-K. Thielemann, M. Wiescher, End point of the *rp* process on accreting neutron stars. Phys. Rev. Lett. **86**, 3471–3474 (2001)

62. J.L. Fisker, H. Schatz, F.-K. Thielemann, Explosive hydrogen burning during type I X-ray bursts. Astrophys. J. Suppl. Ser. **174**, 261 (2008)

63. A. Parikh, J. José, G. Sala, C. Iliadis, Nucleosynthesis in type I X-ray bursts. Prog. Part. Nucl. Phys. **69**, 225–253 (2013)

64. A. Parikh, J. José, C. Iliadis, F. Moreno, T. Rauscher, Impact of uncertainties in reaction Q values on nucleosynthesis in type I X-ray bursts. Phys. Rev. C **79**, 045802 (2009)

65. H. Schatz, The importance of nuclear masses in the astrophysical rp-process. Int. J. Mass Spectrom. **251**, 293–299 (2006)

66. C. Iliadis, *Nuclear Physics of Stars* (Wiley, Hoboken, 2007)

67. H. Schatz, W.-J. Ong, Dependence of X-ray burst models on nuclear masses. Astrophys. J. **844**, 139 (2017)

68. M. Arnould, S. Goriely, K. Takahashi, The r-process of stellar nucleosynthesis: astrophysics and nuclear physics achievements and mysteries. Phys. Rep. **450**, 97–213 (2007)

69. A. Arcones, F.-K. Thielemann, Neutrino-driven wind simulations and nucleosynthesis of heavy elements. J. Phys. G **40**, 013201 (2013)

70. P. Mösta, C.D. Ott, D. Radice, L.F. Roberts, E. Schnetter, R. Haas, A large-scale dynamo and magnetoturbulence in rapidly rotating core-collapse supernovae. Nature **528**, 376 (2015)

71. F.-K. Thielemann, M. Eichler, I. Panov, B. Wehmeyer, Neutron star mergers and nucleosynthesis of heavy elements. Annu. Rev. Nucl. Part. S. **67**, 253–274 (2017)

72. B.P. Abbott et al., Multi-messenger observations of a binary neutron star merger. Astrophys. J. Lett. **848**, L12 (2017)

73. B.P. Abbott et al., GW170817: observation of gravitational waves from a binary neutron star inspiral. Phys. Rev. Lett. **119**, 161101 (2017)

74. A. Goldstein, P. Veres, E. Burns, M.S. Briggs, R. Hamburg, D. Kocevski, C.A. Wilson-Hodge, R.D. Preece, S. Poolakkil, O.J. Roberts, C.M. Hui, V. Connaughton, J. Racusin, A. von Kienlin, T.D. Canton, N. Christensen, T. Littenberg, K. Siellez, L. Blackburn, J. Broida, E. Bissaldi, W.H. Cleveland, M.H. Gibby, M.M. Giles, R.M. Kippen, S. McBreen, J. McEnery, C.A. Meegan, W.S. Paciesas, M. Stanbro, An ordinary short gamma-ray burst with extraordinary implications: Fermi-GBM detection of GRB 170817A. Astro-phys. J. Lett. **848**, L14 (2017)

75. V. Savchenko, C. Ferrigno, E. Kuulkers, A. Bazzano, E. Bozzo, S. Brandt, J. Chenevez, T.J.-L. Courvoisier, R. Diehl, A. Domingo, L. Hanlon, E. Jourdain, A. von Kienlin, P. Laurent, F. Lebrun, A. Lutovinov, A. Martin-Carrillo, S. Mereghetti, L. Natalucci, J. Rodi, J.-P. Roques, R. Sunyaev, P. Ubertini, Integral detection of the first prompt gamma-ray signal coincident with the gravitational-wave event GW170817. Astrophys. J. Lett. **848**, L15 (2017)

76. R. Chornock, E. Berger, D. Kasen, P.S. Cowperthwaite, M. Nicholl, V.A. Villar, K.D. Alexander, P.K. Blanchard, T. Eftekhari, W. Fong, R. Margutti, P.K.G. Williams, J. Annis, D. Brout, D.A. Brown, H.-Y. Chen, M.R. Drout, B. Farr, R.J. Foley, J.A. Frieman, C.L. Fryer, K. Herner, D.E. Holz, R. Kessler, T. Matheson, B.D. Metzger, E. Quataert, A. Rest, M. Sako, D.M. Scolnic, N. Smith, M. Soares-Santos, The electromagnetic counterpart of the binary neutron star merger LIGO/VIRGO GW170817. IV. Detection of near-infrared signatures of r-process nucleosynthesis with Gemini-south. Astrophys. J. Lett. **848**, L19 (2017)

77. E. Pian, P. D'Avanzo, S. Benetti, M. Branchesi, E. Brocato, S. Campana, E. Cappellaro, S. Covino, V. D'Elia, J. Fynbo et al., Spectroscopic identification of r-process nucleosynthesis in a double neutron-star merger. Nature **551**, 67 (2017)

78. M. Mumpower, R. Surman, G. McLaughlin, A. Aprahamian, The impact of individual nuclear properties on r-process nucleosynthesis. Prog. Part. Nucl. Phys. **86**, 86–126 (2016)

79. P. Moller, J. Nix, W. Myers, W. Swiatecki, Nuclear ground-state masses and deformations. At. Data Nucl. Data Tables **59**, 185–381 (1995)

80. M. Liu, N. Wang, Y. Deng, X. Wu, Further improvements on a global nuclear mass model. Phys. Rev. C **84**, 014333 (2011)

81. H. Zhang, J. Dong, N. Ma, G. Royer, J. Li, H. Zhang, An improved nuclear mass formula with a unified prescription for the shell and pairing corrections. Nucl. Phys. A **929**, 38–53 (2014)

82. H. Koura, T. Tachibana, M. Uno, M. Yamada, Nuclidic mass formula on a spherical basis with an improved even-odd term. Progr. Theor. Phys. **113**, 305–325 (2005)

83. J. Duflo, A. Zuker, Microscopic mass formulas. Phys. Rev. C **52**, R23–R27 (1995)

84. M.W. Kirson, An empirical study of the Duflo–Zuker mass formula. Nucl. Phys. A **893**, 27–42 (2012)

85. S. Goriely, F. Tondeur, J. Pearson, A Hartree–Fock nuclear mass table. At. Data Nucl. Data Tables **77**, 311–381 (2001)

86. K.-L. Kratz, K. Farouqi, P. Möller, A high-entropy-wind r-process study based on nuclear-structure quantities from the new finite-range droplet model FRDM (2012). Astrophys. J. **792**, 6 (2014)

87. S. Goriely, N. Chamel, J.M. Pearson, Skyrme-Hartree-Fock-Bogoliubov nuclear mass formulas: crossing the 0.6 MeV accuracy threshold with microscopically deduced pairing. Phys. Rev. Lett. **102**, 152503 (2009)

88. S. Goriely, N. Chamel, J.M. Pearson, Further explorations of Skyrme-Hartree-Fock-Bogoliubov mass formulas. XIII. The 2012 atomic mass evaluation and the symmetry coefficient. Phys. Rev. C **88**, 024308 (2013)

89. M. Wang, G. Audi, A.H. Wapstra, F.G. Kondev, M. MacCormick, X. Xu, B. Pfeiffer, The Ame2012 atomic mass evaluation. Chin. Phys. C **36**, 1603–2014 (2012)

90. D. Atanasov, P. Ascher, K. Blaum, R.B. Cakirli, T.E. Cocolios, S. George, S. Goriely, F. Herfurth, H.-T. Janka, O. Just, M. Kowalska, S. Kreim, D. Kisler, Y.A. Litvinov, D. Lunney, V. Manea, D. Neidherr, M. Rosenbusch, L. Schweikhard, A. Welker, F. Wienholtz, R.N. Wolf, K. Zuber, Precision mass measurements of $^{129-131}$Cd and their impact on stellar nucleosynthesis via the rapid neutron capture process. Phys. Rev. Lett. **115**, 232501 (2015)

91. C. Babcock, R. Klawitter, E. Leistenschneider, D. Lascar, B.R. Barquest, A. Finlay, M. Foster, A.T. Gallant, P. Hunt, B. Kootte, Y. Lan, S.F. Paul, M.L. Phan, M.P. Reiter, B. Schultz, D. Short, C. Andreoiu, M. Brodeur, I. Dillmann, G. Gwinner, A.A. Kwiatkowski, K.G. Leach, J. Dilling, Mass measurements of neutron-rich indium isotopes toward the $N = 82$ shell closure. Phys. Rev. C **97**, 024312 (2018)

92. J. Van Schelt, D. Lascar, G. Savard, J.A. Clark, S. Caldwell, A. Chaudhuri, J. Fallis, J.P. Greene, A.F. Levand, G. Li, K.S. Sharma, M.G. Sternberg, T. Sun, B.J. Zabransky, Mass measurements near the r-process path using the Canadian penning trap mass spectrometer. Phys. Rev. C **85**, 045805 (2012)

93. R. Orford, N. Vassh, J.A. Clark, G.C. McLaughlin, M.R. Mumpower, G. Savard, R. Surman, A. Aprahamian, F. Buchinger, M.T. Burkey, D. A. Gorelov, T.Y. Hirsh, J.W. Klimes, G.E. Morgan, A. Nystrom, K.S. Sharma, Precision mass measurements of neutron-rich neodymium and samarium isotopes and their role in understanding rare-earth peak formation. Phys. Rev. Lett. **120**, 262702 (2018)

94. M. Vilen, J.M. Kelly, A. Kankainen, M. Brodeur, A. Aprahamian, L. Canete, T. Eronen, A. Jokinen, T. Kuta, I.D. Moore, M.R. Mumpower, D.A. Nesterenko, H. Penttilä, I. Pohjalainen, W.S. Porter, S. Rinta-Antila, R. Surman, A. Voss, J. Äystö, Precision mass measurements on neutron-rich rare-earth iso-topes at JYFLTRAP: reduced neutron pairing and implications for r-process calculations. Phys. Rev. Lett. **120**, 262701 (2018)

95. Y.X. Watanabe, Y.H. Kim, S.C. Jeong, Y. Hirayama, N. Imai, H. Ishiyama, H.S. Jung, H. Miyatake, S. Choi, J.S. Song, E. Clement, G. de France, A. Navin, M. Rejmund, C. Schmitt, G. Pollarolo, L. Corradi, E. Fioretto, D. Montanari, M. Niikura, D. Suzuki, H. Nishibata, J. Takatsu, Pathway for the production of neutron-rich isotopes around the $N = 126$ shell closure. Phys. Rev. Lett. **115**, 172503 (2015)

96. C.H. Dasso, G. Pollarolo, A. Winther, Systematics of isotope production with radioactive beams. Phys. Rev. Lett. **73**, 1907–1910 (1994)

97. V. Zagrebaev, W. Greiner, Production of new heavy isotopes in low-energy multinucleon transfer reactions. Phys. Rev. Lett. **101**, 122701 (2008)

98. Y. Watanabe, Y. Hirayama, N. Imai, H. Ishiyama, S. Jeong, H. Miyatake, E. Clement, G. de France, A. Navin, M. Rejmund, C. Schmitt, G. Pollarolo, L. Corradi, E. Fioretto, D. Montanari, S. Choi, Y. Kim, J. Song, M. Niikura, D. Suzuki, H. Nishibata, J. Takatsu, Study of collisions of 136Xe+198Pt for the KEK isotope separator. Nucl. Instr. Meth. Phys. Res. B **317**; XVIth International Conference on Electromagnetic Isotope Separators and Techniques Related to their Applications, December 2–7, 2012 at (Matsue, Japan, 2013), pp. 752–755

99. Y. Hirayama, Y.X. Watanabe, N. Imai, H. Ishiyama, S.C. Jeong, H. Miyatake, M. Oyaizu, M. Mukai, S. Kimura, Y.H. Kim, T. Sonoda, M. Wada, M. Huyse, P.V. Duppen, Beta-decay spectroscopy of r-process nuclei with $N = 126$ at KEK isotope separation system, in Proceedings of the Conference on Advances in Radioactive Isotope Science (aris2014)

100. T. Kurtukian-Nieto, J. Benlliure, K.-H. Schmidt, L. Audouin, F. Becker, B. Blank, E. Casarejos, F. Farget, M. Fernández-Ordóñez, J. Giovinazzo, D. Henzlova, B. Jurado, J. Pereira, O. Yordanov, Production cross sections of heavy neutron-rich nuclei approaching the nucleosynthesis r-process path around $A = 195$. Phys. Rev. C **89**, 024616 (2014)

101. A. Winther, Grazing reactions in collisions between heavy nuclei. Nucl. Phys. A **572**, 191–235 (1994)

102. A. Winther, Dissipation, polarization and fluctuation in grazing heavy-ion collisions and the boundary to the chaotic regime. Nucl. Phys. A **594**, 203–245 (1995)

103. Y. Hirayama, H. Miyatake, Y.X. Watanabe, N. Imai, H. Ishiyama, S.C. Jeong, H.S. Jung, M. Oyaizu, M. Mukai, S. Kimura, T. Sonoda, M. Wada, Y.H. Kim, M. Huyse, Yu. Kudryavtsev, P. Van Duppen, Beta-decay spectroscopy of r-process nuclei around N = 126. EPJ. Web of Conf. **109**, 08001 (2016)

Chapter 2
Half-Life Measurement of ^{11}C for Testing the Standard Model

2.1 Motivation

Among the superallowed mixed mirror decays, ^{11}C is of particular interest due to its importance to the search for physics beyond the Standard Model. If there are additional interactions alongside the vector and axial-vector interactions of $V - A$ theory, they would be present in the calculation as an additional term $\left(1 + \frac{\gamma b_F}{W}\right)$ in the integrand of the statistical rate function. Here, W is the total electron energy in electron rest mass units, $\gamma = \sqrt{1 - (\alpha Z)^2}$, with Z the atomic number of the daughter nucleus and α the fine structure constant, and b_F is the Fierz interference term [1]. The latter is related to the ratio of scalar coupling or tensor coupling to vector coupling or axial-vector coupling, respectively [2]. As the lighter $T = 1/2$ mixed decay nuclei have smaller Q_{EC} values, and thus W values, their decays are most sensitive to physics beyond the Standard Model, though such sensitivity would be limited by the uncertainty on the determination of ρ. ^{11}C is the lightest such nucleus that undergoes β^+ decay. Since ^{11}C decays completely to the ^{11}B ground state, a branching ratio measurement is unnecessary, and a recent high-precision Q_{EC} value measurement [3] has made the half-life the largest remaining source of experimental uncertainty, other than the unmeasured ρ. Hence, a new, higher-precision half-life measurement of ^{11}C was conducted in July 2017 [4].

2.2 Experimental Method

This new ^{11}C measurement was conducted at the University of Notre Dame's Nuclear Science Laboratory (NSL), making use of the FN tandem Van de Graaf accelerator and *TwinSol* mass separator. The location of the setup within the NSL is illustrated in Fig. 2.1.

© Springer Nature Switzerland AG 2019
A. A. Valverde, *Precision Measurements to Test the Standard Model and for Explosive Nuclear Astrophysics*, Springer Theses,
https://doi.org/10.1007/978-3-030-30778-3_2

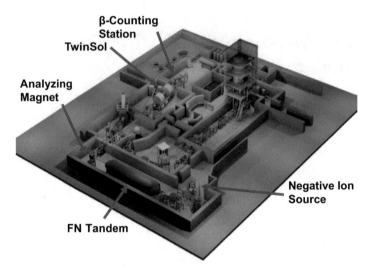

Fig. 2.1 Layout of NSL at Notre Dame, labeling the negative ion source (SNICS), FN tandem accelerator, analyzing magnet, *TwinSol* mass separator, and the β-counting station

2.2.1 FN Tandem

In the production of the ^{11}C, a primary beam of ^{10}B$^-$ was first created using the Source of Negative Ions by Cesium Sputtering (SNICS) ion source and a ^{10}B-Ag cathode. This beam was steered into the FN tandem. In the FN tandem, a pelletron system is used to charge the terminal to a voltage of several megavolts; this accelerates the negatively-charged beam from SNICS into a stripper foil located at the terminal, stripping electrons from the negatively-charged beam and generating a range of positive charge states, which are further accelerated, and then exit the FN tandem into a mass analyzing magnet that is used to select the final charge state of the primary beam. In this experiment, the terminal voltage of the FN tandem was 6.5 MV, and a primary beam of 32.5 MeV ^{10}B^{4+} was selected using the mass analyzing magnet.

2.2.2 TwinSol

The primary ^{10}B^{4+} beam was then impinged on a deuterium gas target, which produced ^{11}C through the ^{10}B$(d, n)^{11}$C reaction. The resulting rare isotope beam was then passed through the *TwinSol* [5] mass separator, which consists of a pair of superconducting solenoids capable of producing a magnetic field of up to 6T. These are used as a mass separator, selecting an 18 MeV ^{11}C^{6+} secondary beam.

Fig. 2.2 The University of
Notre Dame β-counting
station, Fig. 2 from [6].
Labelled are (1) the rotatable
arm, (2) the port through
which the ion beam enters the
station, (3) two holders for
the tantalum foil, only one of
which was used in this
experiment, and (4) the
plastic scintillator connected
to the photomultiplier tube

2.2.3 β-Counting Station

The ^{11}C ions were then implanted in a thick tantalum foil in the NSL's β-
counting station [6, 7] (see Fig. 2.2[1]), which consists of a circular aluminum
chamber containing a rotating aluminum arm on which a tantalum foil was mounted
for implantation. The measurement was then conducted following the procedure
outlined in Refs. [6, 8], with the primary beam turned off during the counting stage
by deflecting it with a high voltage kicker upstream of the FN tandem Van de
Graaff accelerator. ^{11}C was implanted in the tantalum foil for 60 min (approximately
three half-lives), and then the foil was rotated into the counting position and the
decay was measured. The individual betas were counted using a 1 mm plastic
scintillator mounted to a light guide that was cemented to the photomultiplier
tube. The photomultiplier tube used was an ET-Enterprises 9266QKSB, featuring
a quartz window to minimize background from radioactive material and a mu-
metal$^{\circledR}$ shield, mounted to a high-pulse linearity RB1108 base. A thin $(8(2)\,\mu m)$,
light-tight aluminum foil was placed in front of the plastic scintillator; the thickness
of the aluminum foil was minimized to maximize the recorded betas from the ^{11}C
decay, which have a $Q_{\beta+} = 1981.69(6)$ keV [9]. A series of nine implantations and
half-life measurements were conducted in this way, varying the photomultiplier tube
bias, discriminator threshold, and beam current (and thus initial count rate) between
individual measurement runs in order to probe possible systematic effects.

---

[1]Reprinted figure with permission from Brodeur et al. [6]. Copyright 2016 by the American
Physical Society.

2.3 Half-Life Determination

The data analysis follows the procedure previously used in half-life measurements at the University of Notre Dame [6, 8]. The data for each experimental run consisted of a single cycle containing a decay measurement and one or more cycles containing background measurements taken during implantation, which were accounted for by eliminating the runs with fewer than 10% of the average counts of all runs from consideration. Each remaining cycle contained between 1.9 and 11.1 million total detected counts, taken over 220 min or approximately 11 half-lives for the first run, and 380 min or approximately 19 half-lives for the remaining runs. The leading bins were excluded to avoid bins with anomalously low counts, and the data was rebinned to avoid the presence of a large number of empty bins, which could introduce a bias into the fit [8]. The initial 6600 and 11,400 bins were rebinned to 600 bins, which was selected as it optimized the χ^2_ν of the fit.

2.3.1 Poisson Fitting

An important consideration when fitting to determine a half-life measurement is that such a data set comes from a counting experiment. This means that the underlying data set must be fit using a method that is based on Poisson statistics, rather than the common least-squares fit, which is based on a Gaussian or normal distribution. A Gaussian least-squares fit would result in a small systematic shift to the fit half-life [10]; this has been recognized as an important consideration in half-life measurement since at least 1969 [11], and two different Poisson-distribution-based fitting procedures are commonly used.

2.3.2 Iterative Fitting Procedure

The iterative fitting procedure is laid out in Ref. [12] and has been used previously at the University of Notre Dame as the primary fitting procedure [6, 8]. As the final eight runs were of the same cycle length, they could be combined into a single data set and fit as an ensemble; since the first run had a different length, it was considered separately. The first step in the analysis is to adjust the counts in each bin for the dead-time losses inherent in the system; this is done by taking the measured counts per bin and generating the dead-time corrected data for each summed bin i, $\overline{D}(i)$, through [12]:

$$\overline{D}(i) = \sum_n \frac{D_n(i)}{1 - \frac{D_n(i)\tau}{t_{\mathrm{bin}}}} \tag{2.1}$$

where $D_n(i)$ is the number of counts in a given bin in run n, τ is the system dead-time per event, and t_{bin} is the bin width. Also calculated is the dead-time corrected Poisson variance of each bin, $\overline{V}(i)$ [12]:

$$\overline{V}(i) = \sum_n \frac{D_n(i)}{\left(1 - \frac{D_n(i)\tau}{t_{bin}}\right)^2} \tag{2.2}$$

The fit function, $\overline{Y}(i)$, is calculated based on the rate function, $r(t)$. For ^{11}C, there was no observed radioactive contamination, so these were:

$$r(t) = r_0 \exp\left[\frac{-\ln(2)t}{t_{1/2}}\right] + b \tag{2.3}$$

$$\overline{Y}(i) = N \int_{t_{begin}}^{t_{end}} r(t)dt \tag{2.4}$$

with an initial rate r_0, half-life $t_{1/2}$, and background rate b for the rate calculation. The total number of summed runs is N, and the fit function is integrated between bin beginning and ending times t_{begin} and t_{end}. A weighting function $\overline{W}(i)$ is then generated [12]:

$$\frac{1}{\overline{W}(i)} = \overline{V}(i)\frac{\overline{Y}(i)}{\overline{D}(i)} \tag{2.5}$$

The fitting is done using the generated weights $\overline{W}(i)$ and a Levenberg–Marquart least-squares fitting algorithm [12]; the modified weights compensate for the Gaussian assumptions of the fit. Furthermore, the fit is performed iteratively, recalculating (2.4) and (2.5) from the newly fit parameters and rerunning the fit until the relative change in all parameters is less than 0.01%, which usually occurs within fewer than ten iterations. The summed fit and corresponding residuals of the dead-time corrected data for the combined runs 2–9 can be seen in Fig. 2.3.[2] The residuals average -0.004 with a standard deviation of 0.932. The resulting half-life from the summed fit was $t_{1/2} = 1221.38(89)$ s for the first run, and $t_{1/2} = 1220.20(22)$ s for the summed fit of runs 2 through 9. These values are consistent with each other and have a weighted average of $t_{1/2} = 1220.27(22)$ s.

---

[2]Reprinted figure with permission from Brodeur et al. [6]. Copyright 2016 by the American Physical Society.

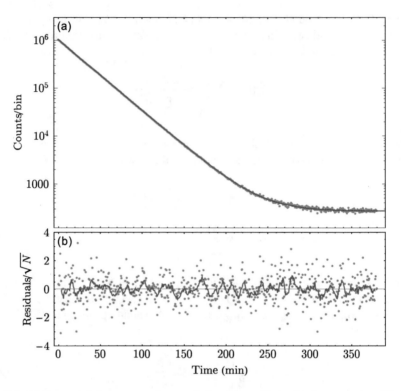

Fig. 2.3 (**a**) Summed β decay curves for runs 2–9 together with the fitted curve (red; solid). (**b**) Residuals divided by the square root of the fit number of counts in a given bin N and a 10-point moving average (red; solid). Each bin is 38 s wide

2.3.3 Poisson χ^2 Fitting Procedure

A second common fitting algorithm [13] was also used to fit this data, serving as a cross-check to the sum fitting algorithm. Here, instead of the classic Pearson's χ^2 [14]

$$\chi^2_{\text{Pearson}} = \sum_i \frac{(y_i - y_{\text{fit}})^2}{y_{\text{fit}}} \tag{2.6}$$

where y_i is the expected number of counts, and y_{fit} is the fit number of counts, a Poisson-statistics-derived χ^2 is used instead [13]

$$\chi^2_{\text{Poisson}} = 2 \sum_i W_i \left(y_{\text{fit}} - y_i + y_i \ln \left[\frac{y_i}{y_{\text{fit}}} \right] \right) \tag{2.7}$$

where the weights are $W_i = \frac{y_i}{\overline{V}(i)}$, with the variance $\overline{V}(i)$ as defined in Eq. (2.2). A Levenberg–Marquart least-squares fitting algorithm is used with this alternate χ^2, again yielding a half-life with minimal bias from the fitting algorithm. This fitting algorithm is more likely to diverge if the initial parameters are not well-chosen, but as a cross-check, it works well. For these data sets, the resulting reduced χ_ν^2 equals 1.04, and the fit half-life is $t_{1/2} = 1221.40(89)$ for the first run and $t_{1/2} = 1220.20(16)$ for runs 2 through 9, consistent with the results using the summed fit algorithm. The weighted difference between the half-lives calculated these two algorithms was 0.004 s, considerably less than the statistical uncertainty, but this was still considered as a systematic uncertainty.

2.4 Uncertainty Determination

Beyond the statistical uncertainty determined from the fitting algorithm, a tabulation of all the systematic uncertainties is critical for the presentation of a high-precision half-life determination [15]. While an inspection of the residuals in panel b of Fig. 2.3 indicates the absence of any non-statistical trend, as illustrated by the 10-point moving average, the presence of contaminants, the uncertainty in the dead time, and several possible other sources of systematic uncertainty were considered.

2.4.1 Contaminant-Related Considerations

The most significant possible source of systematic error in the half-life value comes from the possibility of radioactive contamination. An ion chamber was used to study the composition of the cocktail beam emerging from the *TwinSol* separator. The resulting particle identification plot, Fig. 2.4,[3] shows no radioactive isotopes beyond the ^{11}C, and the heaviest isotopes produced were beryllium, boron, and carbon. Thus, heavier radioactive isotopes of nitrogen or oxygen were not produced and could not be contaminants. The energy of the primary beam was selected such that the production of other radioisotopes via ^{10}B—deuterium reactions was energetically forbidden, with the exception of long-lived ^{7}Be and ^{3}H. The beryllium only decays via electron capture, and the beta decay of tritium is too low energy to pass the aluminum foil in front of the detector. Moreover, the 12 year half-life of tritium would have minimally affected our background.

Nevertheless, fits for the observed decay rate $r(t)$ with two decaying half-lives were conducted, using:

---

[3]Reprinted figure with permission from Valverde et al. [4]. Copyright 2018 by the American Physical Society.

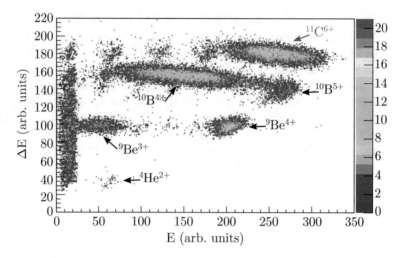

Fig. 2.4 Particle identification plot of the incoming cocktail beam separated by the *TwinSol* facility near the location of the β-counting station, showing energy lost in the first section of the ion chamber vs. residual energy lost in the rest of the ion chamber. Faint periodic groups can be seen alongside the identified isotope groups, which are the result of interactions with the wires of the chamber

$$r(t) = r_0 \left(\exp\left[\frac{-\ln(2)t}{t_{1/2}} \right] + \alpha \exp\left[\frac{-\ln(2)t}{t_2} \right] \right) + b \qquad (2.8)$$

where r_0, $t_{1/2}$, and b are defined as in Eq. (2.3), t_2 is the half-life of the possible contaminant, and α is the contamination ratio. With a free-floating t_2, this fit resulted in $t_2 = 2 \times 10^3 (3 \times 10^7)$ min and $\alpha = 4 \times 10^{-10} (5 \times 10^{-5})$; fixing t_2 as half or double that of ^{11}C result in $\alpha = 3 \times 10^{-9} (3 \times 10^{-3})$ and $\alpha = 6 \times 10^{-10} (2 \times 10^{-4})$, respectively, all of which are consistent with zero. The possibility of a much longer-lived contaminant produced by the activation of the beamline was also investigated. Such an activation is rendered extremely unlikely due to the 18 MeV energy of the secondary beam being below the Coulomb barrier for reactions with the nuclei in the primary components of the stainless steel beamline, though it is possible on the aluminum of the paddle holding the tantalum foil; this would be a very small area exposed to an incident rate of less than 10^4 pps. The beam is turned off during the counting phase to further reduce the dose to the aluminum, and the counting station itself is located 12 m from the production target and separated from it by a 1.5 m thick high-density concrete wall, resulting in an immeasurably small neutron flux. Nevertheless, the possibility for the production of a long-lived contaminant polluting the spectra was probed by adding linear dependence of slope m to our background:

$$r(t) = r_0 \exp\left[-\frac{\ln(2)t}{t_{1/2}} \right] + mt + b \qquad (2.9)$$

where m is the slope from the decay of the very long-lived contaminant. For this last fit, we found a slope $m = -0.002(40)$ s^{-1}, which is consistent with zero.

Possible short-lived contaminants and the possible mis-evaluation of the dead time were studied through removing the leading bins one by one and then performing a summed half-life fit on the remaining data. Up to the first 220 min were removed, corresponding to approximately eleven half-lives and over 99.8% of the total counts; any further removal of data does not result in a meaningful fit. As can be seen in Fig. 2.5,[4] no time-dependent systematic trends are apparent in either the full data set or in the two subsets with varying initial count rates. Thus, it can be safely concluded that there were no radioactive contaminants present in this half-life determination.

2.4.2 Other Systematic Considerations

To search for other possible systematic effects, the photomultiplier tube bias voltage, discriminator threshold voltage, and beam current were all varied. The photomultiplier tube was biased at 1000 and 1100 V, the discriminator set at 0.3 and 0.5 V, and the primary beam current was varied resulting in initial β count rates ranging from 1500 to 10,500 per second. The background varied from 0.3 to 1.6 counts/s on individual runs, depending on the PMT bias and threshold voltage. Combinations of these parameters were explored in each run, and the fitting procedure was performed individually to probe systematics. As can be seen in the top panel of Fig. 2.6,[5] there are no apparent systematic effects; the absence of any rate-dependent effects here or in Fig. 2.5 further indicates that there are no rate-dependent photomultiplier tube gain shifts [16].

The possible statistical nature of the larger spread in the half-life at low initial rates was tested through 100 different Monte Carlo simulated data sets with the same initial rates and background as the experimental data sets. As indicated by the Monte Carlo generated sample data set at the bottom panel of Fig. 2.6, a similar scatter between the experimental and simulated data sets is observed. Furthermore, as indicated by the shaded region on the bottom panel of Fig. 2.6, the one standard deviation spread calculated from the 100 simulated data sets overlaps with the spread in the data.

The weighted average of these individual runs gives a half-life of $t_{1/2} = 1220.24(22)$ s, which is in agreement with the value from the sum fit. The small, 26 ms difference can be explained by a bias of the maximum likelihood fitting with count rate [12] and is replicated in the 36 ms average spread from the 100

[4]Reprinted figure with permission from Valverde et al. [4]. Copyright 2018 by the American Physical Society.

[5]Reprinted figure with permission from Valverde et al. [4]. Copyright 2018 by the American Physical Society.

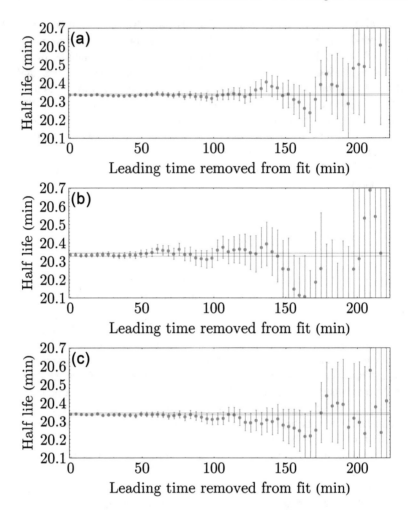

Fig. 2.5 Fitted half-lives for the summed data as a function of the leading time removed. The two horizontal (red) lines are the uncertainty on the half-life without any bin removal. (**a**) Represents the sum of all eight same-length runs, (**b**) the sum of the three runs with an initial count rate of approximately 3000 counts per second, and (**c**) the sum of the two runs with an initial count rate of approximately 10,000 counts per second

Monte Carlo generated data sets. Nevertheless, half of the experimental difference is added as a systematic uncertainty. The Birge ratio [17] or square-root of χ_ν^2 of these measurements, 0.95(16), indicates that the variation in values is statistical in nature. Finally, the accuracy of the iterative fitting algorithm was tested by taking the weighted average of the iterative fits for each of the 100 Monte Carlo generated data sets. The difference of $-11(18)$ ms with the inputted half-life, which is consistent with zero, demonstrates the accuracy of the iterative fitting technique. Nevertheless, to be conservative, an uncertainty of 18 ms is added as a systematic uncertainty.

2.4.2.1 Dead-Time Uncertainty

The uncertainty in the dead time, $\tau = 56.89(9)\,\mu s$, also affects the ^{11}C half-life. Hence, the summed fit was repeated using the two 1σ limits of τ, $\tau = 56.80$ and $\tau = 56.98\,\mu s$, for these data. Half the difference between the weighted averages for these two cases, 0.14 s, was taken as the systematic uncertainty.

2.4.2.2 Clock Frequency Uncertainty

The 100 Hz clock frequency is known to be accurate to within 0.4 mHz. The summed fit was repeated using the two extrema of the clock value, 100.0004 and 99.9996 Hz for the clock frequency. The difference in half-life was found to be on the order of milliseconds; half of this difference, 0.005 s, was added as a systematic uncertainty.

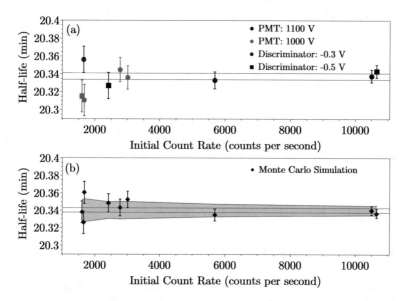

Fig. 2.6 (**a**) Half-life results from fitting individual samples vs initial count rate. Point color indicates the photomultiplier tube voltage, and shape indicates discriminator voltage. The two horizontal (blue) lines are the uncertainty on the weighted average of these values. (**b**) Monte Carlo simulated data of the same varying rates, showing the same statistical scatter around the weighted average half-life, indicated by two horizontal (blue) lines. The gray band indicates the 1σ spread of 100 such simulations

Table 2.1 Summary of statistical and systematic uncertainties combined to give final uncertainty on ^{11}C half-life

Source	Uncertainty (s)
Statistical	0.22
Dead-time uncertainty	0.14
Binning	0.026
Fit (Monte Carlo)	0.018
Fit (individual vs. sum)	0.013
Clock uncertainty	0.005
Fit ([12] vs. [13])	0.004
Total	0.26

2.4.2.3   Binning

The effect of rebinning the data was recently explored using Monte Carlo generated data [8], which showed no systematic effects above the statistical uncertainty provided few bins had zero counts. The difference in half-life between rebinning into 200 and 1000 bins is on the order of hundredths of seconds, but again, half of that difference, 0.026 s, was added as a systematic uncertainty.

2.4.3   Summary

The statistical and systematic uncertainties are summarized in Table 2.1. To produce the final uncertainties, these values were all added in quadrature, producing a total uncertainty of 0.26 s.

2.5   Results

2.5.1   World-Average Half-Life

Our new half-life value, $t_{1/2} = 1220.27(26)$ s, is in good agreement with the previous world average, $t_{1/2}^{old} = 1221.6(1.5)$ s, but is significantly more precise. Following the same procedure used for previous reviews of superallowed mixed mirror decays [18] and superallowed pure Fermi $0^+ \rightarrow 0^+$ decays [11, 19], we reevaluated the world data in order to calculate a new world average half-life. As our new value is significantly more precise than the previous values, seven of those used in the previous evaluation [20–26] were eliminated, being more than ten times less precise, following the established procedure [18, 19]. This leaves four earlier values that are used to calculate the new world average [27–30]. These, alongside our new

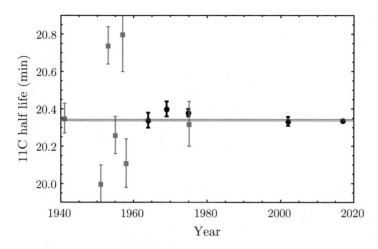

Fig. 2.7 Half-lives of ^{11}C considered in the calculation of the new world average [27–30] (Black circles), as well as those considered previously and now eliminated [20–26] (Gray squares). The horizontal (red) lines represent the uncertainty on $t_{1/2} = 1220.41(32)$ s

measurement, can be seen in Fig. 2.7.[6] A weighted average of the five measurements was taken, giving a world average of $t_{1/2}^{\mathrm{world}} = 1220.41(32)$ s, which is a factor of five more precise than the previous value. The Birge ratio of our new average is 1.28(21), an improvement on the previous value of 2.06(14). As it is greater than one, we adopt the policy of the Particle Data Group [31], and the uncertainty reported on the world average has been scaled by the Birge ratio.

2.5.2 $\mathcal{F}t^{\mathrm{mirror}}$

Using our new world average half-life and the recent value for f_V from [3], we can now calculate a new value for $\mathcal{F}t^{\mathrm{mirror}}$ following Eq. (1.17). A summary of the values used in this calculation and their sources can be seen in Table 2.2. Our new value is an improvement of a factor of 2.6 in the uncertainty over the previous value. This now makes the ^{11}C $\mathcal{F}t^{\mathrm{mirror}}$-value the most precise to date, with a level of precision comparable to the most precise $\mathcal{F}t^{0^+ \to 0^+}$ values.

Table 2.2 Parameters used in calculation of $\mathcal{F}t^{\text{mirror}}$ and related values

Ref.	Parameter	Value
Valverde et al. [4]	$t_{1/2}^{\text{world}}$	1220.41(32) s
Gulyuz et al. [3]	Q_{EC}	1981.690(61) keV
Gulyuz et al. [3]	f_V	3.1829(8)
Valverde et al. [4]	f_A	3.2163(8)
Severijns et al. [18]	δ'_R	1.660(4)%
Severijns et al. [18]	$\delta_C^V - \delta_{NS}^V$	1.04(3)%
Hardy and Towner [11]	$\mathcal{F}t^{0^+ \to 0^+}$	3072.27(72) s

Table 2.3 Comparison of calculated values from Valverde et al. [4] to Gulyuz et al. [3]

Parameter	Valverde et al. [4]	Gulyuz et al. [3]
$\mathcal{F}t^{\text{mirror}}$	3916.9(1.9) s	3920.4(5.0) s
ρ	0.75022(56)	0.7493(15)
a_{SM}	0.51982(46)	0.5206(13)
A_{SM}	$-0.59962(2)$	$-0.59959(5)$
B_{SM}	$-0.8877(3)$	$-0.8872(8)$

2.5.3   Standard Model Estimates of ρ and Correlation Coefficients

This new $\mathcal{F}t^{\text{mirror}}$ value allows us to extract a Standard Model prediction for ρ using the world-average $\mathcal{F}t^{0^+ \to 0^+}$, obtained from the 14 most precise Fermi superallowed $0^+ \to 0^+$ decays [11]. Using $|M_f^0|^2 = 1$ for $T = 1/2$ mirror decays and $|M_f^0|^2 = 2$ for the pure Fermi $T = 1$ decays, we can determine from Eqs. (1.14) and (1.18) [18] that:

$$\mathcal{F}t^{\text{mirror}} = \frac{2\mathcal{F}t^{0^+ \to 0^+}}{1 + \frac{f_A}{f_V}\rho^2} \tag{2.10}$$

where f_A was calculated from the Q_{EC} in Ref. [3] and the parameterization presented in Ref. [32]. This was then solved for ρ, and this value, as well as the correlation coefficients A_{SM}, a_{SM}, and B_{SM}, was calculated following the Standard Model [18]. As in Ref. [18], our calculated correlation coefficients at $E_\beta = 0$ include neither physics beyond the Standard Model nor recoil order effects, which would affect measured correlation coefficients. These results can be seen in Table 2.3, resulting in significant reduction on the uncertainty by factors of three to five over the previous results.

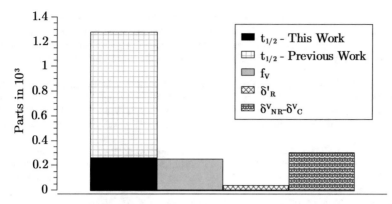

Fig. 2.8 Fractional contribution of experimental and theoretical parameters to the final uncertainty in $\mathcal{F}t^{\text{mirror}}$

2.5.4 Fractional Contributions to Uncertainty

Having improved the precision of the measured world-average half-life of ^{11}C, the lightest and longest-lived superallowed mixed mirror β^+ decay, by a factor of five, it is now comparable in precision to the Q value and thus increases the precision of the $\mathcal{F}t^{\text{mirror}}$ value fourfold. In examining the fractional contributions to the final uncertainty of the $\mathcal{F}t^{\text{mirror}}$ value, Fig. 2.8,[7] we can see that the largest uncertainty now comes from the theoretical $\delta^V_{NS} - \delta^V_C$ correction, providing an impetus for improved precision calculations. The new estimate for ρ and the related Standard Model correlation coefficients also show a significant improvement over the previous results. The high precision achieved on the $\mathcal{F}t^{\text{mirror}}$ value is now the most accurate of any superallowed mixed mirror decay and is comparable in precision to the most precise $\mathcal{F}t^{0^+ \to 0^+}$ values. With this measurement, it would only take a relative precision of 0.5% on a measurement of ρ to determine V_{ud} with a relative uncertainty of 0.2%, comparable to the uncertainties on Fermi superallowed decays that currently provide the most precise determinations of V_{ud} [11].

References

1. J. Jackson, S. Treiman, H. Wyld, Coulomb corrections in allowed beta transitions. Nucl. Phys. **4**, 206–212 (1957)
2. J.C. Hardy, I.S. Towner, Superallowed $0^+ \to 0^+$ nuclear β decays: a new survey with precision tests of the conserved vector current hypothesis and the standard model. Phys. Rev. C **79**, 055502 (2009)

---

[7]Reprinted figure with permission from Valverde et al. [4]. Copyright 2018 by the American Physical Society.

3. K. Gulyuz, G. Bollen, M. Brodeur, R.A. Bryce, K. Cooper, M. Eibach, C. Izzo, E. Kwan, K. Manukyan, D.J. Morrissey, O. Naviliat-Cuncic, M. Redshaw, R. Ringle, R. Sandler, S. Schwarz, C.S. Sumithrarachchi, A.A. Valverde, A.C.C. Villari, High precision determination of the β decay Q_{EC} value of ^{11}C and implications on the tests of the standard model. Phys. Rev. Lett. **116**, 012501 (2016)

4. A.A. Valverde, M. Brodeur, T. Ahn, J. Allen, D.W. Bardayan, F.D. Becchetti, D. Blankstein, G. Brown, D.P. Burdette, B. Frentz, G. Gilardy, M.R. Hall, S. King, J.J. Kolata, J. Long, K.T. Macon, A. Nelson, P.D. O'Malley, M. Skulski, S.Y. Strauss, B. Vande Kolk, Precision half-life measurement of ^{11}C: the most precise mirror transition Ft value. Phys. Rev. C **97**, 035503 (2018)

5. F. Becchetti, J. Kolata, Recent results from the TwinSol low-energy rib facility. Nucl. Instrum. Meth. Phys. Res. B **376**, 397–401 (2016). Proceedings of the XVIIth International Conference on Electromagnetic Isotope Separators and Related Topics (EMIS2015), Grand Rapids, MI, 11–15 May 2015

6. M. Brodeur, C. Nicoloff, T. Ahn, J. Allen, D.W. Bardayan, F.D. Becchetti, Y.K. Gupta, M.R. Hall, O. Hall, J. Hu, J.M. Kelly, J.J. Kolata, J. Long, P. O'Malley, B.E. Schultz, Precision half-life measurement of ^{17}F. Phys. Rev. C **93**, 025503 (2016)

7. T. Ahn, D. Bardayan, D. Bazin, S.B. Novo, F. Becchetti, J. Bradt, M. Brodeur, L. Carpenter, Z. Chajecki, M. Cortesi, A. Fritsch, M. Hall, O. Hall, L. Jensen, J. Kolata, W. Lynch, W. Mittig, P. O'Malley, D. Suzuki, The proto-type active-target time-projection chamber used with TwinSol radioactive-ion beams. Nucl. Instrum. Meth. Phys. Res. B **376**, 321–325 (2016). Proceedings of the XVIIth International Conference on Electromagnetic Isotope Separators and Related Topics (EMIS2015), Grand Rapids, MI, 11–15 May 2015

8. J. Long, T. Ahn, J. Allen, D.W. Bardayan, F.D. Becchetti, D. Blankstein, M. Brodeur, D. Burdette, B. Frentz, M.R. Hall, J.M. Kelly, J.J. Kolata, P.D. O'Malley, B.E. Schultz, S.Y. Strauss, A.A. Valverde, Precision half-life measurement of ^{25}Al. Phys. Rev. C **96**, 015502 (2017)

9. M. Wang, G. Audi, F. Kondev, W. Huang, S. Naimi, X. Xu, The AME2016 atomic mass evaluation (II). Tables, graphs and references. Chin. Phys. C **41**, 030003 (2017)

10. J.M. Freeman, J. Jenkin, G. Murray, D. Robinson, W. Burcham, The ft value of the superallowed Fermi beta decay 26mal(β+)26mg. Nucl. Phys. A **132**, 593–610 (1969)

11. J.C. Hardy, I.S. Towner, Superallowed $0^+ \rightarrow 0^+$ nuclear β decays: 2014 critical survey, with precise results for V_{ud} and CKM unitarity. Phys. Rev. C **91**, 025501 (2015)

12. V. Koslowsky, E. Hagberg, J. Hardy, G. Savard, H. Schmeing, K. Sharma, X. Sun, The half-lives of ^{42}Sc, ^{46}V, ^{50}Mn and ^{54}Co. Nucl. Instr. Meth. Phys. Res. A **401**, 289–298 (1997)

13. G.F. Grinyer, C.E. Svensson, C. Andreoiu, A.N. Andreyev, R.A.E. Austin, G.C. Ball, R.S. Chakrawarthy, P. Finlay, P.E. Garrett, G. Hackman, J.C. Hardy, B. Hyland, V.E. Iacob, K.A. Koopmans, W.D. Kulp, J.R. Leslie, J.A. Macdonald, A.C. Morton, W.E. Ormand, C.J. Osborne, C.J. Pearson, A.A. Phillips, F. Sarazin, M.A. Schumaker, H.C. Scraggs, J. Schwarzenberg, M.B. Smith, J.J. Valiente-Dobón, J.C. Waddington, J.L. Wood, E.F. Zganjar, High precision measurements of ^{26}Na β^- decay. Phys. Rev. C **71**, 044309 (2005)

14. S. Baker, R.D. Cousins, Clarification of the use of chi-square and likelihood functions in fits to histograms. Nucl. Instr. Meth. Phys. Res. **221**, 437–442 (1984)

15. Referee A, Report of Referee A – LH16325/Valverde, LH16325 later published as [4] (2017)

16. S. Triambak, P. Finlay, C.S. Sumithrarachchi, G. Hackman, G.C. Ball, P.E. Garrett, C.E. Svensson, D.S. Cross, A.B. Garnsworthy, R. Kshetri, J.N. Orce, M.R. Pearson, E.R. Tardiff, H. Al-Falou, R.A.E. Austin, R. Churchman, M.K. Djongolov, R. D'Entremont, C. Kierans, L. Milovanovic, S. O'Hagan, S. Reeve, S.K.L. Sjue, S.J. Williams, High-precision measurement of the ^{19}Ne half-life and implications for right-handed weak currents. Phys. Rev. Lett. **109**, 042301 (2012)

17. R.T. Birge, The calculation of errors by the method of least squares. Phys. Rev. **40**, 207–227 (1932)

18. N. Severijns, M. Tandecki, T. Phalet, I.S. Towner, Ft values of the $T = 1/2$ mirror β transitions. Phys. Rev. C **78**, 055501 (2008)

19. J.C. Hardy, I.S. Towner, Superallowed $0^+ \to 0^+$ nuclear β decays: a critical survey with tests of the conserved vector current hypothesis and the standard model. Phys. Rev. C **71**, 055501 (2005)

20. J.H.C. Smith, D.B. Cowie, The measurement of artificial radioactivity in liquid tracer samples using c11. J. Appl. Phys. **12**, 78–82 (1941)

21. J.M. Dickson, T.C. Randle, The excitation function for the production of 7 be by the bombardment of 12 c by protons. Proc. Phys. Soc. A **64**, 902 (1951)

22. D.N. Kundu, T.W. Donaven, M.L. Pool, J.K. Long, Nuclear reactions with 21-MeV He^3 in a cyclotron. Phys. Rev. **89**, 1200–1202 (1953)

23. W.C. Barber, W.D. George, D.D. Reagan, Absolute cross section for the reaction $C^{12}(\gamma, n)C^{11}$. Phys. Rev. **98**, 73–76 (1955)

24. I.D. Prokoshkin, A. Tiapkin, Investigation of the excitation functions for the reactions $C^{12}(p,pn)C^{11}$, $Al^{27}(p,3pn)Na^{24}$ and $Al^{27}(p,3p,3n)Na^{22}$ in the 150–660 MeV energy range. Sov. Phys. JETP **5**, 148 (1957)

25. S. Arnell, J. Dubois, O. Almén, Half-lives of the positron emitting mirror nuclides. Nucl. Phys. **6**, 196–202 (1958)

26. H. Behrens, M. Kobelt, L. Szybisz, W.-G. Thies, On the $^{11}C \to {}^{11}B \, \beta^+$ transition. Nucl. Phys. A **246**, 317–322 (1975).

27. T.M. Kavanagh, J.K.P. Lee, W.T. Link, A precise measurement of the C11 production cross section for 98-MeV protons on carbon. Can. J. Phys. **42**, 1429–1436 (1964)

28. M. Awschalom, F. Larsen, W. Schimmerling, Activation of air near a target bombarded by 3 GeV protons. Nucl. Instr. Meth. **75**, 93–102 (1969)

29. G. Azuelos, J.E. Kitching, Half-lives of some $T = \frac{1}{2}$ mirror decays. Phys. Rev. C **12**, 563–566 (1975)

30. D. Woods, M. Baker, J. Keightley, L. Keightley, J. Makepeace, A. Pearce, A. Woodman, M. Woods, S. Woods, S. Waters, Standardisation of 11C. Appl. Radiat. Isot. **56**, 327–330 (2002). Proceedings of the Conference on Radionuclide Metrology and its Applications, ICRM'01

31. J. Beringer, J.-F. Arguin, R.M. Barnett, K. Copic, O. Dahl, D.E. Groom, C.-J. Lin, J. Lys, H. Murayama, C.G. Wohl, W.-M. Yao et al., Review of particle physics. Phys. Rev. D **86**, 010001 (2012)

32. I.S. Towner, J.C. Hardy, Parametrization of the statistical rate function for select superallowed transitions. Phys. Rev. C **91**, 015501 (2015)

Chapter 3
The LEBIT Facility and Penning Traps

3.1 Radioactive Beam Production at the NSCL

The Low-Energy Beam and Ion Trap (LEBIT) facility [1] is located at Michigan State University's National Superconducting Cyclotron Laboratory (NSCL). The LEBIT facility is unique among Penning trap mass spectrometry facilities in its ability to perform high-precision mass measurements on rare isotopes produced through projectile fragmentation. This is made possible by the rare isotope production and in-flight separation facilities at the NSCL, which enable the production of high intensity rare isotope beams. The principal equipment employed for production and separation is the coupled cyclotron facility (CCF).

Beam production in the CCF begins with the production of a beam of heavy, highly charged stable ions using an Electron Cyclotron Resonance (ECR) source, which are then injected into the smaller K500 cyclotron. The K500 then accelerates the ion beam to $\sim$14 MeV/u, and the beam is then extracted and injected into the larger K1200 cyclotron. In the K1200, the remaining electrons are removed using a stripper foil, and the beam is accelerated to $\sim$140 MeV/u. The primary beam is then extracted and focused on the production target, a thin, light target commonly composed of beryllium that is used for the production of the rare isotope secondary beam. The secondary beam retains most of the initial beam energy after the target. The fragments are separated using the A1900 fragment separator [2], which uses an energy-degrading wedge and two dispersive beamline section to achieve general isotopic separation of the secondary beam. This secondary beam can then be delivered to one of the several experimental vaults; measurements using the LEBIT facility require delivery of the beam to the gas stopping area in the N4 vault.

© Springer Nature Switzerland AG 2019
A. A. Valverde, *Precision Measurements to Test the Standard Model and for Explosive Nuclear Astrophysics*, Springer Theses,
https://doi.org/10.1007/978-3-030-30778-3_3

3.2   Gas Stopping Area

The projectile fragmentation method of producing rare isotope beams employed at the NSCL produces high-energy, high-emittance beams; however, Penning trap mass spectrometry and other low energy experiments require low-energy, low-emittance beams. At the NSCL, this conversion of beam properties is done in the gas stopping area, located in the N4 vault after the A1900 fragment separator.

Beams enter the gas stopping area through a momentum-compression beamline, where a series of solid (usually aluminum) degraders and a final monochromatic wedge are used, respectively, to reduce the energy of the beam to less than 1 MeV/u and reduce the beam's energy spread to allow for more efficient stopping of the beam in the following gas cell. The gas cell is a large-volume RF-based gas catcher constructed by Argonne National Labs [3]. It consists of a 1.2 m long volume of gas, in which the ions are thermalized through collisions with high-purity helium buffer gas, usually at ~100 mbar. The high ionization potential of helium forces the ions to remain ionized in the 1+ or 2+ charge states. As the gas flow through the chamber is relatively slow, an electrostatic drift field on the order of tens of V/cm is used to sweep the positive ions towards the extraction electrodes. The walls of the gas cell and a cone located at the extraction end are composed of closely spaced RF electrodes that repel the injected and thermalized ions and guide them to extraction. A radiofrequency quadrupole (RFQ) ion guide [4] is used in the extraction optics to guide the ions through three stages of differential pumping before they are reaccelerated to 30 keV/q and sent on to a mass separating magnet and then the d-line to the low-energy area. Figures 3.1[1] and 3.2 show a schematic diagram of the structure of the gas cell, and an image of the gas cell as installed at the NSCL.

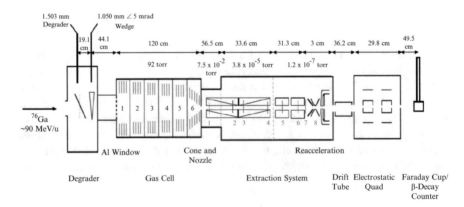

Fig. 3.1 Figure 1 from [3], showing a schematic layout of the gas cell as used in the commissioning experiment. Distances between components are given in cm (not to scale)

---

[1]Reprinted from Cooper et al. [3], Copyright 2014, with permission from Elsevier.

Fig. 3.2 Photograph of the gas cell installed at the NSCL

3.3 Offline Ion Sources

In addition to rare isotope beam from the gas stopping area, LEBIT can also take beam from two offline production sources. The first offline source installed in LEBIT is a modified commercial ion source from the Colutron Research Corporation, which can produce stable isotopes in two ways, either through surface ionization or from a plasma. When operated as a surface ionization source, the filament in the Colutron source is heated, vaporizing either the alkali impurities present in the filament itself or an alkali earth metal loaded into the ceramic holder and inserted into the center of the heated filament coil. Once vaporized, the alkali or alkali earth metal vapor is ionized by contact with the hot filament, which is biased to ∼100 V, which causes the ions to accelerate away from the filament. These are then extracted and accelerated into the LEBIT cooler-buncher. This source has been used in this way to produce alkali earth isotopes like ^{48}Ca [5, 6] for the study of neutrinoless double β decay, and the alkali isotopes produced from the filament (commonly ^{39}K, ^{41}K, ^{85}Rb, and ^{133}Cs) have been used as either reference ions for rare isotope measurements (e.g., as in Ref. [7]) or for the calibration of mass-dependent shifts in Penning trap mass measurements [8]. For use as a plasma ion source, a gas is injected into the chamber through a needle valve and the filament is negatively biased to ∼ − 100 V, causing it to emit electrons. At a pressure of ∼10^{-6} mbar, a continuous discharge occurs, ionizing the gas and generating a plasma. This mode of operation can also be used to produce ions of elements with low melting points, where a plasma is generated using helium gas, and the element of interest is vaporized in a ceramic holder and then ionized by the plasma. The source has been used in this way to produce ions such as ^{82}Se [9] for neutrinoless double β decay experiments and for the production of ^{14}N in the direct measurement of the ^{14}O Q_{EC} value [10].

The second offline source at LEBIT is a laser ablation source, or LAS [11, 12]. Here, a Quantel Brilliant pulsed neodymium-doped yttrium aluminum garnet (Nd:YAG) laser with a second harmonic module produces 532 nm light with a pulse duration of 4 ns. The power is usually limited to well below its maximum, commonly achieving a density of $\sim$5 $\times$ 10^8 W/cm^2. This laser is focused on a solid, rotating target, where the laser pulse vaporizes target material, and the high temperature of the target caused by the laser irradiation results in the emission of positive ions and electron, resulting in a multi-stage plasma expansion [11]. This has been used to produce ionized clusters of ^{12}C ions, which are extremely useful as a calibrant due to the definition of the atomic mass unit, and for the production of ions of ^{50}Ti, ^{50}V, ^{50}Cr [13] ^{96}Zr [8], ^{113}Cd [14], and ^{190}Pt [15], all for the study of neutrinoless double β decay, or for the production of ^{11}B for the measurement of the ^{11}C Q_{EC} value [8], among others.

3.4   Beam Cooler and Buncher

Beam in the LEBIT facility comes from gas stopping area or either of the two offline ion sources; a 90° bender is used to steer beam from the offline sources while deflecting beam from the other sources, or disabled to allow beam from the gas stopping area to enter the LEBIT beam cooler and buncher [16]. The LEBIT cooler-buncher is a two-stage gas-filled RFQ ion trap that accepts continuous beam and converts it to low-energy, low-emittance ion pulses for efficient ion capture upon injection into the LEBIT Penning trap [17]. Figure 3.3 shows the two sections

Fig. 3.3 Photographs of the LEBIT cooler-buncher, showing (**a**) the cooler section and μRFQ, and (**b**) the buncher section and ejection optics. Ceramic mug for scale

of the LEBIT cooler-buncher. The first RFQ stage uses a moderate buffer gas, usually ~0.02 mbar helium, to "cool" or slow the ions and reduce their transverse emittance through collisions in the large diameter RFQ ion guide which provides an electrostatic gradient dragging the ions towards the second section. The second stage is operated at a lower pressure (usually ~10^{-4} mbar), and consists of a segmented RFQ which is used to generate a trapping potential to perform the final cooling and accumulation of ions before the bunched ions are released to the LEBIT Penning trap in pulses of approximately 100 ns [18]. A lower pressure in this second section is desirable to minimize beam reheating during extraction. A small-diameter μRFQ provides efficient transport between these two sections while also allowing differential pumping.

3.5 9.4 T Penning Trap

Mass measurements at LEBIT are currently performed using the 9.4 T Penning trap mass spectrometer [18, 19], though a second trap, the Single-Ion Penning Trap (SIPT) is currently being commissioned. Trapping in a Penning trap is illustrated in the schematic in Fig. 3.4,[2] where a homogeneous magnetic field created by the 9.4 T superconducting magnet provides radial confinement, while the hyperbolic

Fig. 3.4 Schematic diagram of an ideal Penning trap, Fig. 2a from [18], with hyperbolic ring and endcap electrodes generating an electrostatic quadrupole potential for longitudinal confinement and a magnetic field for radial confinement

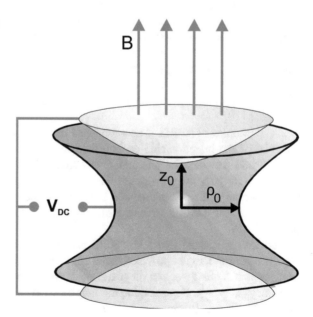

---

[2]Reprinted from Ringle et al. [18], Copyright 2009, with permission from Elsevier.

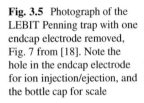

Fig. 3.5 Photograph of the LEBIT Penning trap with one endcap electrode removed, Fig. 7 from [18]. Note the hole in the endcap electrode for ion injection/ejection, and the bottle cap for scale

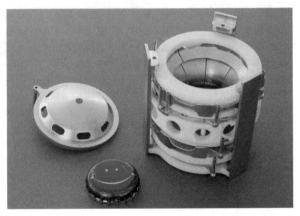

electrode structure (seen in Fig. 3.5[3]) creates an electrostatic quadrupole potential that provides axial confinement.

3.5.1   Trapping

The quadrupole potential is generated from a potential difference of $V_0 \sim 25$ V between the ring and endcap electrodes of the LEBIT Penning trap. As shown in Fig. 3.4, these are both hyperbola, with the ring electrode having an inner radius of $\rho_0 = 13$ mm, and the spacing between the two endcaps of $2z_0 = 22.36$ mm. Additional correction ring and tube electrodes exist to correct for deviations from a pure quadrupole potential due to the holes in the endcaps for injection and ejection of ions, and the finite size of the electrodes; the trap was designed to minimize such deviations [19].

Axial trapping at LEBIT is accomplished through a dynamic method, where the potential of the injection-side endcap is switched rapidly. When an ion bunch is injected into the trap, the potential is lowered, allowing it to enter; once the ion bunch is in the trap, the potential of this electrode is rapidly raised, trapping the ion bunch. Timing of the rise in potential is optimized such that the ion bunch is located in the middle of the trap when the injection electrode is changed to the trapping potential [19].

The magnetic field B of the 9.4 T actively shielded, persistent, solenoidal, superconducting magnet provides radial confinement for the ions, as a moving ion of mass m and charge q in a magnetic field will undergo cyclotron motion with a frequency $\omega_c = \frac{qB}{m}$. Superconducting magnet based on Ni$_3$Sn technology such as the LEBIT magnet is known to have relatively large but very linear magnetic field

---

[3]Reprinted from Ringle et al. [18], Copyright 2009, with permission from Elsevier.

decay rates [20]. The LEBIT 9.4 T magnet's decay has been measured as $\frac{\Delta B}{B}/\Delta t \sim$ -8×10^{-8}/h, and is compensated for by running current through a pair of insulated copper wires wound around the bore tube of the magnet [18, 21]. Furthermore, the pressure of the helium bath is stabilized using a precision barometer and a valve controlled with a PID loop; variations in pressure affect the evaporation of the liquid helium in the bath, and thus cause nonlinear changes in the magnetic field strength.

3.5.2   Ion Eigenmotions

An ion captured in a Penning trap undergoes a characteristic motion, which has been solved exactly for an ideal Penning trap [22]. This motion is a superposition of three independent eigenmotions, as seen in Fig. 3.6.[4] These are the axial (z), magnetron (−), and reduced cyclotron (+) motions, each of which has an eigenfrequency, ω_z, ω_-, and ω_+. These are most clearly defined based on two additional parameters, the previously discussed true angular cyclotron frequency $\omega_c = \frac{qB}{m}$ and the characteristic trap parameter d. The characteristic trap parameter is defined in terms of the trap dimensions ρ_0 and z_0 as $d = \sqrt{\frac{z_0^2}{2} + \frac{\rho_0^2}{4}}$. The eigenfrequencies are then

$$\omega_z = \sqrt{\frac{qV_0}{md^2}} \tag{3.1}$$

$$\omega_\pm = \frac{\omega_c}{2} \pm \sqrt{\frac{\omega_c^2}{4} - \frac{\omega_z^2}{2}} \tag{3.2}$$

where q, m, and V_0 are the charge and mass of the trapped ion and the trapping potential. For a singly charged ion of $m = 50$ u in the LEBIT 9.4 T Penning trap,

Fig. 3.6 Motions of an ion in a Penning trap, Fig. 2b from [18]. This motion is a superposition of the three eigenmotions, axial, magnetron, and reduced cyclotron

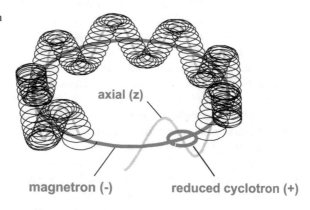

axial (z)

magnetron (-)   reduced cyclotron (+)

---

[4]Reprinted from Ringle et al. [18], Copyright 2009, with permission from Elsevier.

these frequencies are roughly $\nu_+ = 3\,\mathrm{MHz}$, $\nu_z = 100\,\mathrm{kHz}$, and $\nu_- = 2\,\mathrm{kHz}$, and in general, $\nu_+ > \nu_z > \nu_-$.

There are three additional important relations among the frequencies, the first of which is called the invariance theorem [23].

$$\omega_c^2 = \omega_+^2 + \omega_-^2 + \omega_z^2 \tag{3.3}$$

$$\omega_c = \omega_+ + \omega_- \tag{3.4}$$

$$\omega_z^2 = 2\omega_+\omega_- \tag{3.5}$$

While the invariance theorem is true even for small ($\sim 10^{-3}$) misalignments with the magnetic field and distortions of the trapping potential [22, 24], the TOF-ICR technique makes use of Eq. (3.5).

3.6 Ion Preparation

Ion bunches released from the cooler-buncher are rarely isotopically pure; these impurities come from many sources, including the original cocktail beam, the ionization and chemical recombination of impurities in the gas cell, and the decay of radioactive ions of interest. However, the presence of more than one ion species in the Penning trap during a mass measurement can lead to systematic errors in the determined mass [25], due to the Coulomb interaction of the ion of interest with the contaminant ions. Thus, a multi-stage system of beam purification has been implemented at LEBIT [26]. The first stage occurs on extraction from the gas cell, where the magnetic dipole mass separator generally achieves a resolving power greater than 500, separating non-isobaric contaminants.

The second stage occurs on extraction from the cooler-buncher; a fast kicker following the extraction optics [27] provides a time-of-flight mass filter. This is done using a fast switching deflector, which is composed of pairs of plates in the horizontal and vertical directions. A voltage is applied to prevent ions from reaching the Penning trap. When the isotope of interest passes, a fast ($\sim 300\,\mathrm{ns}$) switch drops the deflector voltage to zero, allowing transport to the trap. As the pulses are of $\sim 100\,\mathrm{ns}$ FWHM, this gives a resolving power greater than 400. This is necessary to remove non-isobaric contaminants from ions produced in the offline sources, which do not pass through the magnetic dipole mass separator, and to remove any contamination produced in the cooler-buncher through, e.g., charge exchange with contaminant molecules or radioactive decay.

Following the fast kicker, the ion bunch passes through the injection optics, a series of cylindrical electrodes. The final electrode is quadrisected radially, forming a "Lorentz steerer" [19, 28]. This uses an electrostatic bias across opposing sections to create an electric dipole field, which in combination with the magnetic field of the $9.4\,\mathrm{T}$ solenoid results in an $\mathbf{E} \times \mathbf{B}$ force on the ion, thus causing off-axis injection into the Penning trap. This is done to provide initial magnetron (ω_-) motion of the ion, which is necessary for the time-of-flight ion cyclotron resonance (TOF-ICR) technique used at LEBIT that will be discussed in the next section. The final stage of purification, which removes isobaric contaminants, occurs in the Penning trap itself.

3.6.1 Dipole Excitation

The motion of a trapped ion can be excited through the application of multipolar RF fields. At LEBIT, this is done through the segmented ring electrode, where RF voltages can be applied to each individual segment. As an 2^n-pole excitation requires 2^n segments, this means that the eightfold segmented LEBIT ring electrode can support up to octopole excitation [19]. An electric dipole excitation at an eigenfrequency will result in an increase of the amplitude of the associated eigenmotion. Thus, dipole excitation is an effective method of cleaning contaminant ions out of the trap [29, 30]. An excitation at the reduced cyclotron ($\omega_+ = 2\pi\nu_+$) eigenfrequency will drive the ion to a large radius and out of the trap or to such a radius that it cannot be extracted. In the trap, a dipole excitation is generated by applying opposite-phase RF to opposing sections of the ring electrode, as illustrated in the left panel of Fig. 3.7.

3.6.2 Narrowband Dipole Cleaning

The use of dipole excitation of the reduced cyclotron eigenfrequency for cleaning contaminant ions out of the trap relies on the fact that the reduced cyclotron frequency is mass-dependent. By applying dipole excitation at the mass-specific

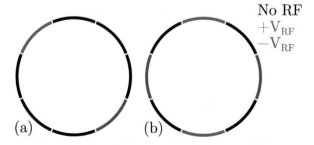

Fig. 3.7 Schematic diagram of LEBIT Penning trap eightfold segmented ring electrode, showing segment usage during (**a**) dipole and (**b**) quadrupole excitations. Black indicates segments to which no voltage is applied, and the RF voltage is applied 180° out of phase between the red ($+V_{RF}$) and blue ($-V_{RF}$) segments

frequencies of known contaminants in the trap, these contaminants can thus be cleaned out of the trap without interfering with the isotope of interest. This relies on identifying the contaminants in the trap; while commonly several contaminants can be easily identified based on the known composition of the beam, it is also common for molecular ions from the gas cell to be present in the bunched beam, which require more effort and time to identify.

3.6.3 SWIFT

The second form of dipole cleaning used at LEBIT instead identifies ranges of frequencies surrounding but excluding a range around the ion of interest to excite. This cleans out all contaminants in these mass bands while leaving the ion of interest unaffected, thus significantly reducing the time that needs to be spent identifying contaminants and allowing more experimental time to be used to conduct mass measurements. The technique used to do this is the Stored Waveform Inverse Fourier Transform (SWIFT) technique [31–34]. The profile of the SWIFT excitation is created in the frequency domain, as illustrated in the top of Fig. 3.8,[5] with two rectangular excitations surrounding an unexcited gap around the frequency ν_+ of the ion of interest. This then undergoes an inverse Fourier transform, producing the time-domain waveform that will be generated by an arbitrary function generator and applied to the Penning trap, as shown in the bottom of Fig. 3.8. As this technique cannot be used when the ion needing to be cleaned is close in mass to the ion of interest (generally the gap must be hundreds of Hz wide to avoid unwanted RF signal), SWIFT and narrowband dipole cleaning are often used in concert.

3.7 Cyclotron Frequency and Mass Determination

After purification, the trapped bunch is then excited and the time of flight to the multi-channel plate (MCP) at the end of the beamline is measured. This allows the determination of the cyclotron frequency ω_c of the ion, and thus the final determination of its mass.

3.7.1 Quadrupole Excitation

If, instead of the opposite phases of RF applied to two opposing plates of the ring electrode, opposite phases are applied in an alternating fashion to two pairs of

---

[5]Reprinted from Kwiatkowski et al. [31], Copyright 2015, with permission from Elsevier.

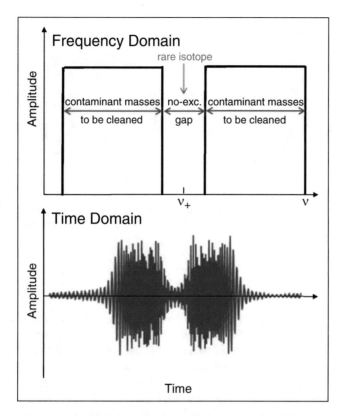

Fig. 3.8 Excitation profiles for SWIFT beam purification at LEBIT, Fig. 1 from [31]. (Top) in the frequency domain, two uniform rectangular excitations are separated by a gap of non-excitation centered on the isotope to be measured. Ions whose ν_+ lies in the excitation bands will be cleaned from the Penning trap. (Bottom) The real part of the inverse Fourier transform of the above frequency-domain signal, which is applied to the electrodes for cleaning

opposed ring segments (illustrated in the right panel of Fig. 3.7), this will create an RF quadrupole excitation, which couples to both radial eigenmotions. If an ion with some initial magnetron motion (created either through off-axis injection via a "Lorentz steerer," as at LEBIT, or through dipole excitation) is excited at $\nu_{RF} = \nu_c = \nu_+ + \nu_-$, this causes the conversion between magnetron and reduced cyclotron motion [25], as illustrated in Fig. 3.9. As $\nu_+ \gg \nu_-$, the conversion from magnetron to reduced cyclotron motion is accompanied by an increase in the energy of the radial motion of the trapped ion.

a)   b)

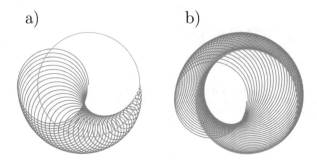

Fig. 3.9 Illustration of the conversion of magnetron to reduced cyclotron motion. (**a**) Ion begins with pure magnetron motion (orange) with amplitude $\rho_- = \rho_0$ and $\rho_+ = 0$, until the quadrupolar RF at $\nu_{RF} = \nu_c$ is applied and the reduced cyclotron radius begins to grow and consequently the magnetron radius decreases. (**b**) Time has passed and full conversion has occurred, with $\rho_- = 0$ and $\rho_+ = \rho_0$

3.7.2 TOF-ICR

The change in the energy of the radial motion is detected using the time-of-flight ion cyclotron resonance (TOF-ICR) technique [35–37]. Following excitation, the voltage on the ejection endcap is lowered, and the ion travels through the ejection optics to an MCP in the Daly configuration [38], where it is detected, and the time of flight relative to the ejection pulse is recorded. When the ion is ejected, its reduced cyclotron motion gives it a magnetic moment $\mu = E_r/B\hat{z}$, where E_r is the energy of the radial motion of the ion, and the magnetic field will generally reduce in strength along the ions path to the MCP. The interaction between the ions magnetic moment and the field gradient results in an axial force, $F = -\mu\frac{\partial B}{\partial z} = \frac{E_r}{B_0} \cdot \frac{\partial B}{\partial z}$, where B_0 is the maximum field strength. When the ion has left the magnetic field, all of the kinetic energy in the radial motion gained in the excitation has been converted to an acceleration along the axis; thus, when the kinetic energy associated with the radial motion is maximized by an excitation at ω_c, the time of flight is minimized.

The total time of flight can be calculated by

$$T(\omega_{RF}) = \int_{z_0}^{z_1} dz \left(\frac{m}{2[E_0 - q \cdot V(z) - \mu(\omega_{RF}) \cdot B(z)]} \right)^{1/2} \tag{3.6}$$

where E_0 is the total initial energy of the ion, q its charge, and $V(z)$ and $B(z)$ are the electric potential and magnetic field strength along the ions path z from the trap (z_0) to the detector (z_1). An ion cyclotron resonance curve can thus be obtained by varying ν_{RF} around ν_c, as seen in Fig. 3.10, and then the actual cyclotron frequency is determined by fitting the theoretical line shape [36] to the data.

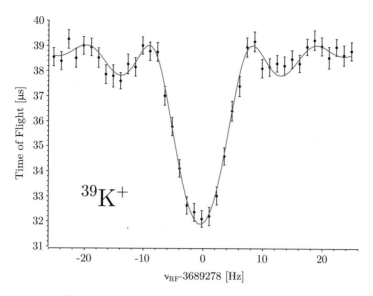

Fig. 3.10 A sample $^{39}K^+$ time-of-flight ion cyclotron resonance with an excitation time $T_{RF} = 100$ ms. The solid red curve represents a fit of the theoretical profile [36]

3.7.3 Mass Determination

In order to determine the mass using Penning trap mass spectrometry, the magnetic field must also be measured very precisely. This is done using the same TOF-ICR technique to measure the mass of a (usually stable) ion or molecular ion of a very well-known mass, called the reference ion. To maximize precision, TOF-ICR measurements of the ion of interest are interleaved with measurements of the reference ion. As the magnetic field of a persistent superconducting magnet decays over time, the two reference measurements bracketing an ion of interest measurement are used to interpolate the magnetic field strength during the mass measurement, as illustrated in Fig. 3.11.

The primary experimental result of a Penning trap mass measurement is thus $R = \frac{q \cdot \nu_{c,ref}^{int}}{q_{ref} \cdot \nu_c}$, where ν_c is the cyclotron frequency of the ion of interest, $\nu_{c,ref}^{int}$ is the linearly interpolated cyclotron frequency of the reference ion, and q and q_{ref} are their charge states. This can be used to determine the mass of the nuclei of interest m as

$$m = R[m_{ref} - q_{ref} \cdot m_e] + q \cdot m_e \qquad (3.7)$$

where m_{ref} is the reference mass, and m_e is the mass of the electron. Usually, R is replaced with $\bar{R}$, the weighted average of a series of individual measurements R. The electron ionization energies and any applicable molecular binding energies

Fig. 3.11 Illustration of linear interpolation of reference measurements to determine cyclotron frequency of the reference ion at the time of the measurement of cyclotron frequency ratio of the ion of interest. The ratio R of these two cyclotron frequencies can be used to determine the mass of the ion of interest if the mass of the reference ion is known

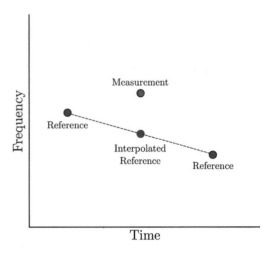

are on the order of eVs, orders of magnitude smaller than the involved masses, are usually too small to have any effect on the statistical uncertainty, and are not included; additional systematic uncertainties will be considered for the specific case presented in the following chapter.

References

1. R. Ringle, G. Bollen, S. Schwarz, Penning trap mass spectrometry of rare isotopes produced via projectile fragmentation at the LEBIT facility. Int. J. Mass Spectrom. **349–350**, 87–93 (2013)
2. D.J. Morrissey, B.M. Sherrill, M. Steiner, A. Stolz, I. Wiedenhoever, Commissioning the A1900 projectile fragment separator. Nucl. Instrum. Methods Phys. Res. Sect. B **204**, 90–96 (2003)
3. K. Cooper, C.S. Sumithrarachchi, D.J. Morrissey, A. Levand, J.A. Rodriguez, G. Savard, S. Schwarz, B. Zabransky, Extraction of thermalized projectile fragments from a large volume gas cell. Nucl. Instrum. Methods Phys. Res. A **763**, 543–546 (2014)
4. B.R. Barquest, An advanced ion guide for beam cooling and bunching for collinear laser spectroscopy of rare isotopes. Ph.D. Thesis, Michigan State University, 2014
5. S. Bustabad, G. Bollen, M. Brodeur, D.L. Lincoln, S.J. Novario, M. Redshaw, R. Ringle, S. Schwarz, A.A. Valverde, First direct determination of the ^{48}Ca double-β decay Q value. Phys. Rev. C **88**, 022501 (2013)
6. M. Redshaw, G. Bollen, M. Brodeur, S. Bustabad, D.L. Lincoln, S.J. Novario, R. Ringle, S. Schwarz, Atomic mass and double-β-decay Q value of ^{48}Ca. Phys. Rev. C **86**, 041306 (2012)
7. A.A. Valverde, G. Bollen, K. Cooper, M. Eibach, K. Gulyuz, C. Izzo, D.J. Morrissey, R. Ringle, R. Sandler, S. Schwarz, C.S. Sumithrarachchi, A.C.C. Villari, Penning trap mass measurement of ^{72}Br. Phys. Rev. C **91**, 037301 (2015)
8. K. Gulyuz, J. Ariche, G. Bollen, S. Bustabad, M. Eibach, C. Izzo, S.J. Novario, M. Redshaw, R. Ringle, R. Sandler, S. Schwarz, A.A. Valverde, Determination of the direct double-β-decay Q value of ^{96}Zr and atomic masses of $^{90-92,94,96}$Zr and $^{92,94-98,100}$Mo. Phys. Rev. C **91**, 055501 (2015)
9. D.L. Lincoln, J.D. Holt, G. Bollen, M. Brodeur, S. Bustabad, J. Engel, S.J. Novario, M. Redshaw, R. Ringle, S. Schwarz, First direct double-β-decay Q-value measurement of ^{82}Se in support of understanding the nature of the neutrino. Phys. Rev. Lett. **110**, 012501 (2013)

10. A.A. Valverde, G. Bollen, M. Brodeur, R.A. Bryce, K. Cooper, M. Eibach, K. Gulyuz, C. Izzo, D.J. Morrissey, M. Redshaw, R. Ringle, R. Sandler, S. Schwarz, C.S. Sumithrarachchi, A.C.C. Villari, First direct determination of the superallowed β-decay Q_{EC} value for ^{14}O. Phys. Rev. Lett. **114**, 232502 (2015)

11. C. Izzo, G. Bollen, S. Bustabad, M. Eibach, K. Gulyuz, D. Morrissey, M. Red-shaw, R. Ringle, R. Sandler, S. Schwarz, A. Valverde, A laser ablation source for offline ion production at LEBIT. Nucl. Instr. Meth. Phys. Res. B **376**, 60–63 (2016). Proceedings of the XVIIth International Conference on Electromagnetic Isotope Separators and Related Topics (EMIS2015), Grand Rapids, MI, 11–15 May 2015

12. S.E. Bustabad, From fundamental fullerenes to the cardinal calcium candidate: the development of a laser ablation ion source and its diverse application at the LEBIT facility. Ph.D. Thesis, Michigan State University, 2014

13. R.M.E.B. Kandegedara, G. Bollen, M. Eibach, N.D. Gamage, K. Gulyuz, C. Izzo, M. Redshaw, R. Ringle, R. Sandler, A.A. Valverde, β-decay Q values among the $A = 50$ Ti-V-Cr isobaric triplet and atomic masses of ^{46,47,49,50}ti, ^{50,51}v, and $^{50,52--54}$Cr. Phys. Rev. C **96**, 044321 (2017)

14. N.D. Gamage, G. Bollen, M. Eibach, K. Gulyuz, C. Izzo, R.M.E.B. Kandegedara, M. Redshaw, R. Ringle, R. Sandler, A.A. Valverde, Precise determination of the ^{113}Cd fourth-forbidden non-unique β-decay Q value. Phys. Rev. C **94**, 025505 (2016)

15. M. Eibach, G. Bollen, M. Brodeur, K. Cooper, K. Gulyuz, C. Izzo, D.J. Morrissey, M. Redshaw, R. Ringle, R. Sandler, S. Schwarz, C.S. Sumithrarachchi, A.A. Valverde, A.C.C. Villari, Determination of the Q_{EC} values of the $T = 1/2$ mirror nuclei ^{21}Na and ^{29}P at LEBIT. Phys. Rev. C **92**, 045502 (2015)

16. T. Sun, High precision mass measurement of ^{37}Ca and developments for LEBIT. Ph.D. Thesis, Michigan State University, 2006

17. S. Schwarz, G. Bollen, R. Ringle, J. Savory, P. Schury, The LEBIT ion cooler and buncher. Nucl. Instr. Meth. Phys. Res. A **816**, 131–141 (2016)

18. R. Ringle, G. Bollen, A. Prinke, J. Savory, P. Schury, S. Schwarz, T. Sun, The LEBIT 9.4 T Penning trap mass spectrometer. Nucl. Instrum. Methods Phys. Res. Sect. A **604**, 536–547 (2009)

19. R. Ringle, P. Schury, T. Sun, G. Bollen, D. Davies, J. Huikari, E. Kwan, D. Morrissey, A. Prinke, J. Savory, S. Schwarz, C.S. Sumithrarachchi, Precision mass measurements with LEBIT at MSU. Int. J. Mass Spectrom. **251**, 300–306 (2006)

20. P. McIntyre, Y. Wu, G. Liang, C.R. Meitzler, Study of Nb$_3$Sn superconducting joints for very high magnetic field NMR spectrometers. IEEE Trans. Appl. Supercond. **5**, 238–241 (1995)

21. R. Ringle, G. Bollen, D. Lawton, P Schury, S. Schwarz, T. Sun, The LEBIT 9.4 T Penning trap system. Eur. Phys. J. A **25**, 59–60 (2005)

22. L.S. Brown, G. Gabrielse, Geonium theory: physics of a single electron or ion in a Penning trap. Rev. Mod. Phys. **58**, 233–311 (1986)

23. L.S. Brown, G. Gabrielse, Precision spectroscopy of a charged particle in an imperfect Penning trap. Phys. Rev. A **25**, 2423–2425 (1982)

24. G. Gabrielse, The true cyclotron frequency for particles and ions in a Penning trap. Int. J. Mass Spectrom. **279**, 107–112 (2009)

25. G. Bollen, H.-J. Kluge, M. König, T. Otto, G. Savard, H. Stolzenberg, R.B. Moore, G. Rouleau, G. Audi, I. Collaboration, Resolution of nuclear ground and isometric states by a Penning trap mass spectrometer. Phys. Rev. C **46**, R2140–R2143 (1992)

26. P. Schury, High precision mass measurements near $N = Z = 33$. Ph.D. Thesis, Michigan State University, 2007

27. J. Savory, High-precision mass measurement of $N \simeq Z \simeq 34$ nuclides for rp-process simulations and developments for the LEBIT facility. Ph.D. Thesis, Michigan State University, 2009

28. R. Ringle, T. Sun, G. Bollen, D. Davies, M. Facina, J. Huikari, E. Kwan, D.J. Morrissey, A. Prinke, J. Savory, P. Schury, S. Schwarz, C.S. Sumithrarachchi, High-precision Penning trap mass measurements of ^{37,38}Ca and their contributions to conserved vector current and isobaric mass multiplet equation. Phys. Rev. C **75**, 055503 (2007)

29. G. Bollen, S. Becker, H.-J. Kluge, M. König, R. Moore, T. Otto, H. Raimbault-Hartmann, G. Savard, L. Schweikhard, H. Stolzenberg, ISOLTRAP: a tandem Penning trap system for accurate on-line mass determination of short-lived isotopes. Nucl. Instr. Meth. Phys. Res. A **368**, 675–697 (1996)
30. K. Blaum, High-accuracy mass spectrometry with stored ions. Phys. Rep. **425**, 1–78 (2006)
31. A.A. Kwiatkowski, G. Bollen, M. Redshaw, R. Ringle, S. Schwarz, Iso-baric beam purification for high precision Penning trap mass spectrometry of radioactive isotope beams with swift. Int. J. Mass Spectrom. **379**, 9–15 (2015)
32. A.A. Kwiatkowski, High-precision mass measurements of ^{32}si and developments at the LEBIT facility. Ph.D. Thesis, Michigan State University, 2011
33. A.G. Marshall, T.C.L. Wang, T.L. Ricca, Tailored excitation for Fourier transform ion cyclotron mass spectrometry. J. Am. Chem. Soc. **107**, 7893–7897 (1985)
34. S. Guan, A.G. Marshall, Stored waveform inverse Fourier transform (swift) ion excitation in trapped ion mass spectrometry: theory and applications. Int. J. Mass Spectrom. Ion Process. **157**, 5–37 (1996)
35. G. Bollen, R.B. Moore, G. Savard, H. Stolzenberg, The accuracy of heavy-ion mass measurements using time of flight-ion cyclotron resonance in a Penning trap. J. Appl. Phys. **68**, 4355–4374 (1990)
36. M. König, G. Bollen, H.-J. Kluge, T. Otto, J. Szerypo, Quadrupole excitation of stored ion motion at the true cyclotron frequency. Int. J. Mass Spectrom. **142**, 95–116 (1995)
37. G. Gräff, H. Kalinowsky, J. Traut, A direct determination of the proton electron mass ratio. Z. Phys. A **297**, 35–39 (1980)
38. N.R. Daly, Scintillation type mass spectrometer ion detector. Rev. Sci. Instrum. **31**, 264–267 (1960)

Chapter 4
Mass Measurement of ^{56}Cu for the Astrophysical *rp*-Process

4.1 Motivation

As discussed previously, reaction pathways around waiting point nuclei are critical to understanding the reaction flow of the astrophysical *rp*-process. With a small Q value for the ^{56}Ni(p, γ) reaction of $Q_{p,\gamma} = 690.3(4)$ keV [1] and an hour-long stellar half-life [2], the doubly magic nucleus ^{56}Ni is one of the most important *rp*-process waiting points [3]. Indeed, it was historically thought to be the endpoint of the *rp*-process [4], though we now know it can proceed to higher masses [5, 6]. The flow through ^{56}Ni is well-characterized, based on Q values [1, 3], as well as ^{56}Ni(p,γ) [7] and ^{57}Cu(p,γ) [8] reaction rates. A route starting at ^{55}Ni could allow *rp*-process flow to bypass the ^{56}Ni waiting point through ^{55}Ni(p,γ)^{56}Cu(p,γ)^{57}Zn(β^+)^{57}Cu but it is not as well-characterized; the branching of the flow at ^{55}Ni between the two routes is determined by the β^+ decay rate and the ^{55}Ni(p,γ) and ^{56}Cu(γ,p) reaction rates. These two different reaction paths are illustrated in Fig. 4.1.

Recently, the low-lying level scheme of ^{56}Cu was experimentally determined for the first time [9], leaving the largest source of uncertainty in the critical ^{55}Ni(p,γ) rate, which can be approximated by Eq. (1.20) to be the proton separation energy of ^{56}Cu. Because of its high astrophysical importance, several predictions of the ^{56}Cu atomic mass have been made recently using the Coulomb displacement energy (CDE) mass relation [10], and the isobaric mass multiplet equation (IMME) [9]. Furthermore, the atomic mass evaluation (AME) predictions varied by several hundreds of keV from AME2003 [11] to AME2012 [12]. Moreover, a precision of better than 10 keV for masses of *rp*-process nuclei is desirable for reliable reaction network calculations [13], a precision which is not achieved by any of the current predictions. The recently released AME2016 includes an unpublished atomic mass from a private communication with Zhang et al. [1] which also fails to achieve the necessary precision. Hence, a high-precision mass measurement of ^{56}Cu

© Springer Nature Switzerland AG 2019
A. A. Valverde, *Precision Measurements to Test the Standard Model and for Explosive Nuclear Astrophysics*, Springer Theses,
https://doi.org/10.1007/978-3-030-30778-3_4

Fig. 4.1 Diagram of the reaction pathways around ^{56}Ni; the primary pathway through ^{56}Ni is in black, and the bypass studied with this mass measurement is in red

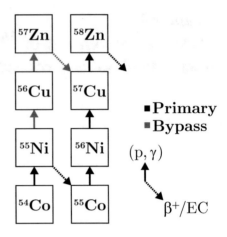

using Penning trap mass spectrometry, the most accurate available technique, was performed at the Low-Energy Beam and Ion Trap (LEBIT) facility at the National Superconducting Cyclotron Laboratory (NSCL) in June 2017 [14].

4.2   Production of ^{56}Cu

The LEBIT facility [15] performs mass measurements on rare isotopes produced via particle fragmentation at the NSCL, as discussed in the previous chapter. In this experiment, radioactive ^{56}Cu was produced by impinging a 160 MeV/u primary beam of ^{58}Ni on a 752 mg/cm^2 beryllium target at the Coupled Cyclotron Facility at the NSCL. The resulting beam passed through the A1900 fragment separator with a 294 mg/cm^2 aluminum wedge [16] to separate the secondary beam. This beam consisted of ^{56}Cu (2.6%), with contaminants of ^{55}Ni, ^{54}Co, and ^{53}Mn.

The beam then entered the beam stopping area [17] through a momentum-compression beamline, where it was degraded with aluminum degraders of 205 μm and 523 μm thickness before passing through a 1010 μm, 3.1 mrad aluminum wedge and entering the gas cell with an energy of less than 1 MeV/u. As previously discussed, the ions are stopped and recombine down to lower charge states through interactions in the gas cell, then extracted through a radiofrequency quadrupole (RFQ) ion guide and transported through a magnetic dipole mass separator with a resolving power greater than 500. Transmitted activity after the mass filter was measured using an insertable Si detector. The most activity was found with $A/q = 92$, corresponding to the extraction of ^{56}Cu as an adduct with two water molecules, $[^{56}\mathrm{Cu(H_2O)_2}]^+$. Following the mass separator, the ions then entered the LEBIT facility; Fig. 4.2 shows a schematic of the path of the beam from the gas stopping area through the LEBIT facility.

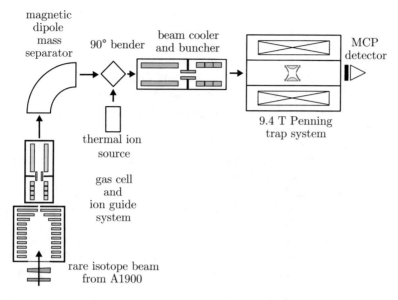

Fig. 4.2 Schematic diagram of the gas stopping area and LEBIT facility

In the LEBIT facility, the $\left[^{56}Cu(H_2O)_2\right]^+$ ions were first injected into the cooler-buncher [18]. On top of its usual operation to cool and bunch the beam, it was also operated to increase the likelihood of molecular breaking following the technique previously used at LEBIT [19]. A potential difference of 55 V between the gas cell and the cooler-buncher accelerated the ions into the helium gas to strip the water ligands using collision-induced dissociation. In this process, molecular ions collide with the buffer gas and generally emit a neutral molecule; this requires the energy of the collision be high enough ($>10\,eV$) to break the molecular bond. The ions were then accumulated, cooled, and released to the LEBIT Penning trap in pulses of approximately 100 ns [20]; the fast kicker in the beam line between the cooler-buncher and the Penning trap was used as a time-of-flight mass separator to select ions of $A/q = 56$, corresponding to $^{56}Cu^+$ and unwanted molecular contaminants of the same A/q. After their capture, the ions were purified, using both dipole cleaning [21] and the stored waveform inverse Fourier transform (SWIFT) technique [22].

4.3   Mass Measurement

Once the ^{56}Cu ions were trapped and other ions were cleaned out of the trap, the previously discussed time-of-flight ion cyclotron resonance technique (TOF-ICR) [23, 24] was used to determine the ions' cyclotron frequency. In these measurements, either a 50, 75, or 100 ms quadrupole excitation was used. These resonances were then fitted to the theoretical line shape [24], and the cyclotron frequency was thus determined; a sample 50-ms resonance of ^{56}Cu$^+$ can be seen in Fig. 4.3.[1] Between measurements of the ^{56}Cu$^+$ cyclotron frequency, measurements of the reference molecular ion C$_4$H$_7^+$ cyclotron frequency were conducted. The C$_4$H$_7$ molecule is possibly the result of an $A = 92$ hydrocarbon molecule extracted from the gas cell and coming with the $[^{56}$Cu(H$_2$O)$_2]^+$ molecule broken by collision-induced dissociation [19].

In Penning trap mass spectrometry, the experimental result is the frequency ratio $R = v_{\text{ref}}^{\text{int}}/v_c$, where $v_{\text{ref}}^{\text{int}}$ is the interpolated cyclotron frequency from the C$_4$H$_7^+$ measurements bracketing the ^{56}Cu$^+$ measurements. A series of 17 measurements of the ^{56}Cu$^+$ cyclotron frequency were taken over a 40-h period and the weighted average of these measurements is $\overline{R} = 1.01641577(12)$. As seen in Fig. 4.4[2] and the Birge ratio [25] of 1.11(12) the individual values of R scatter statistically about the average $\overline{R}$, though the reported uncertainty was scaled by the Birge ratio as it was greater than one, following the policy of the Particle Data Group [26]. Then, using the average of multiple frequency ratios $\overline{R}$ the atomic mass $M(^{56}$Cu$)$ is given by

$$M(^{56}\text{Cu}) = \overline{R}\,[M(\text{C}_4\text{H}_7) - m_e] + m_e \qquad (4.1)$$

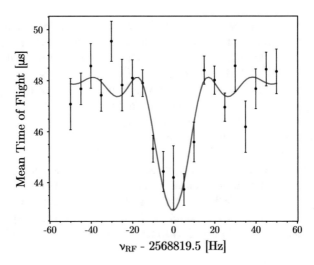

Fig. 4.3 A sample 50-ms ^{56}Cu$^+$ time-of-flight ion cyclotron resonance used for the determination of the frequency ratio of $v_{\text{ref}}^{\text{int}}(\text{C}_4\text{H}_7^+)/v_c(^{56}\text{Cu}^+)$. The solid red curve represents a fit of the theoretical profile [24]

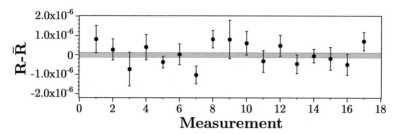

Fig. 4.4 Measured cyclotron frequency ratios $R = v_{\text{ref}}^{\text{int}}/v_c(^{56}\text{Cu}^+)$ relative to the average value $\overline{R}$; the gray bar represents the 1σ uncertainty in $\overline{R}$

where $M(C_4H_7)$ is the mass of the neutral C_4H_7 molecule, and m_e the electron mass. The electron ionization energies and the molecular binding energy of C_4H_7, all on the order of eVs, were not included as they are several orders of magnitude smaller than the statistical uncertainty of the measurement.

4.4 Error Analysis

Most systematic uncertainties in the measured frequency ratios scale linearly with the mass difference between the ion of interest and the calibrant ion. These systematic effects include: magnetic field inhomogeneities, trap misalignment with the magnetic field, harmonic distortion of the electric potential, and non-harmonic imperfections in the trapping potential [23]. All these effects result in a mass-independent shift in the cyclotron frequency ratio; however, the frequency ratio of two ions of different mass depends on their mass. Thus, the measured frequency $R_{\text{meas.}}$ ratio for an ion of interest and its reference ion will differ from the ideal $R_{\text{ideal}} = \frac{v_{\text{ref}}}{v_c}$ as [27]:

$$R_{\text{meas.}} = \frac{v_{\text{ref}} + \Delta v_{\text{ref}}}{v_c + \Delta v_c} \qquad (4.2)$$

where Δv_{ref} and Δv_c are the shifts to the reference ion and ion of interest, respectively. The large, MHz-range value of the cyclotron frequency and the small, Hz-range value of frequency shifts mean that $\delta v_c/v_c \ll 1$, and thus:

$$\frac{\Delta R}{R} = \frac{R_{\text{meas.}} - R_{\text{ideal}}}{R_{\text{ideal}}} \approx \frac{\Delta v_{\text{ref}}}{v_{\text{ref}}} - \frac{\Delta v_c}{v_c} \qquad (4.3)$$

When these frequency shifts are very similar, $\Delta v_{\text{ref}} \approx \Delta v_c = \Delta v$, the frequency shift will have the form $\Delta R/R = (2\pi \Delta v/b)\Delta(m/q)$, where $\Delta(m/q) = (m_{\text{ref}}/q_{\text{ref}} - m/q$ [27]. Mass-dependent shifts to R have been studied at LEBIT through a series of measurements of the well-known masses of ^{38}K,^{85}Rb, and ^{133}Cs as well as several carbon clusters $^{12}C_n$ and found to be at the level of

$2\pi\,\Delta\nu/(m/q) = 2 \times 10^{-10}/(u/e)$ [28, 29]. For the mass shift of 1 u between two same-charge ions present in this measurement, this is negligible compared to the statistical uncertainty on $\overline{R}$.

Remaining systematic effects include nonlinear time-dependent changes in the magnetic field, relativistic effects on the cyclotron frequency, and ion–ion interaction in the trap. Previous work has shown that the effect of nonlinear magnetic field fluctuations on the ratio R should be less than 1×10^{-9} over an hour [30], which was our measurement time. Relativistic effects on R were found to be negligible ($\approx 2 \times 10^{-11}$) due to the large mass of the ions involved.

4.4.1   z-Class Analysis

Isobaric contaminants present in the trap during a measurement could lead to a systematic frequency shift [31]; this effect was minimized by removing most of the contamination using the SWIFT and dipole excitations and by limiting the total number of ions in the trap. For ^{56}Cu, the incident rate limited detected ions in the trap to two or fewer. The number of $C_4H_7^+$ ions was limited by only analyzing events with five or fewer detected ions, which corresponds to 8 or fewer detected ions based on the measured 63% efficiency of the LEBIT MCP [32]. A so-called z-class analysis was performed, where each resonance was divided into the resonances formed by only events with 1, 2, 3, 4, or 5 ions, and these resonances were then fit independently. The weighted average cyclotron frequency ratio R was then calculated for each ion count class. Any count-dependent shifts to R were found to be more than an order of magnitude smaller than the statistical uncertainty.

4.4.2   Systematic Testing of the SWIFT Technique

Possible systematics arising from the use of the SWIFT cleaning technique were probed through a measurement of the ratio R of stable potassium isotopes; $R = \nu_{\text{ref}}^{\text{int}}(^{39}\text{K}^+)/\nu_c(^{41}\text{K}^+)$, with SWIFT being used on the ^{41}K measurement but not for the ^{39}K reference, as in the experiment. Potassium was produced using the LEBIT offline thermal ion source and otherwise treated in the same way as the ions produced online. The measured $\overline{R}$ value agrees with the accepted ratio to within a Birge ratio [25] scaled uncertainty smaller than 2×10^{-8}; individual R values can be seen in Fig. 4.5.[3] Thus, any mass-dependent shifts either from the usage of SWIFT or the difference in mass are negligible compared to the statistical uncertainty on the ^{56}Cu measurement.

---

[3]Reprinted figure with permission from Valverde et al. [14]. Copyright 2018 by the American Physical Society.

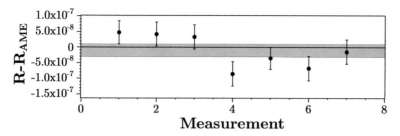

Fig. 4.5 Difference of measured R values of ^{41}K relative to the value calculated from AME2016 [1]. The gray bar represents the average R value and its 1σ uncertainty; the uncertainty of the AME2016 value, 1.5×10^{-10}, is not visible on this graph

Table 4.1 A comparison of predicted mass excesses for ^{56}Cu and $Q_{(p,\gamma)}(^{55}$Ni) with the recent measurement in Valverde et al. [14] and the weighted average of the two experimental measurements

Ref.	ME (keV)	$Q_{(p,\gamma)}(^{55}$Ni) (keV)
Valverde et al. [14]	$-38626.7(7.1)$	579.8(7.1)
AME2016 [1]	$-38643(15)$	596(15)
Experimental average	$-38629.6(6.4)$	582.8(6.4)
Ong et al. [9]	$-38685(82)$	639(82)
Tu et al. [10]	$-38697(88)$	651(88)
AME2003 [11]	$-38600(140)$	560(140)
AME2012 [12]	$-38240(200)$	190(200)

4.5 Results

The resulting mass excess is reported in Table 4.1 as well as the recommended value from the two previous atomic mass evaluations [11, 12], Coulomb displacement energy [10], and the Isobaric Mass Multiplet Equation [9] predictions and the latest result from AME2016 [1]. A comparison can also be seen in Fig. 4.6. Our new ^{56}Cu mass results in $Q_{(p,\gamma)}(^{55}$Ni) $= 579.8(7.1)$ keV, calculated from $Q_{(p,\gamma)}(^{55}$Ni) $= \left[-M(^{56}$Cu$) + M(^{55}$Ni$) + M(^1$H$) \right] c^2$ using our new ^{56}Cu mass and the masses of ^{55}Ni and ^{1}H from AME2016 [1].

4.5.1 Reaction Rate

Using the weighted average of our new ^{56}Cu mass and the AME16 value, also available in Table 4.1, and the level scheme and uncertainties established in Ref. [9], a new astrophysical reaction rate for ^{55}Ni(p, γ) was calculated by Wei Jia Ong following the method outlined in Ref. [9]. The proton and γ widths, Γ_p and Γ_γ were calculated for each state using a shell model with the GXPF1A interaction [33]. Up to three-particle-three-hole excitations in the pf shell were allowed in this calculation, with the proton and γ widths and uncertainties and resonance strengths

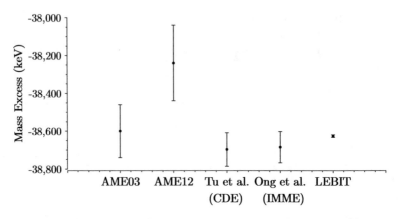

Fig. 4.6 Comparison of the mass excesses reported in AME2003 [11] and AME2012 [12], the CDE [10] and IMME [9] calculated values, and the Penning trap mass measurement [14]

scaled appropriately. A Monte Carlo approach, similar to that in [9, 34], was used to calculate reaction rate uncertainties. At a given temperature, the 16th and 84th percentiles give the $\pm 1\sigma$ uncertainties, and the 49th percentile was used as the median to counter the effects of a skewed distribution from a close-lying resonance. Direct capture rates were calculated using $S(0) = 30.21$ MeV b [35]. Reverse rates are calculated from detailed balance and are most strongly sensitive to the Q-value of the reaction; thus, the reverse rate uncertainty for each Q-value is small and the uncertainty due to the variation of resonance parameters is encompassed within the thickness of the reverse rate line [36]. The results can be seen in Fig. 4.7,[4] compared with the results found using the extrema of the calculated ^{56}Cu masses, AME2012 [12] and Tu et al. [10]; this shows that the (p,γ) reaction dominates up to ≈ 0.3 GK, slightly lower than the Tu et al. case, and significantly higher than the AME2012 case, where the reverse rate always dominates. For the AME2012 mass, at low temperatures, direct capture dominates, leading to little uncertainty, but at higher temperatures, the reaction can access resonant states and the mass uncertainty dominates. Our mass shows a reduced reaction rate uncertainty when compared to these cases, as the Q value uncertainty is now comparable to that in the excitation energy of the resonant states.

4.5.2   Mass Abundance in Ashes

A single-zone X-ray burst model calculation was performed by Wei Jia Ong using the new ^{56}Cu mass with an ignition temperature of 0.386 GK, ignition pressure of 1.73×10^{22} erg cm^{-3}, and initial hydrogen and helium mass fractions of 0.51 and

---

[4]Reprinted figure with permission from Valverde et al. [14]. Copyright 2018 by the American Physical Society.

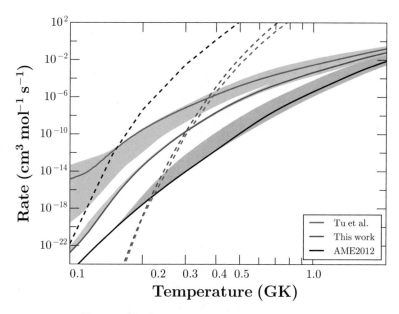

Fig. 4.7 Rate for the ^{55}Ni(p, γ)^{56}Cu reaction and 1σ uncertainties for AME2012 (black band) and Tu et al. masses (red band) and using the experimental mass (blue band). The prior (dashed lines) and new reverse rates (dashed blue line) are also shown

0.39, respectively, demonstrated by Cyburt et al. [37] to produce light curves and ash compositions to most closely match those of multi-zone models, and with a peak temperature of 1.17 GK. As can be seen in Fig. 4.8,[5] the final abundances produced by this calculation demonstrate the extent to which the bypass due to the change in $(p, \gamma) - (\gamma, p)$ equilibrium is active, showing a reduction in abundance in the mass range around the ^{56}Ni waiting point in comparison to ones based on the suggested AME2012 value, though not as extreme as the one seen with the mass from Tu et al.; our maximal bypass is 39%, with a typical X-ray burst trajectory having a bypass of 15%. This means the newly calculated reaction rate allows some of the *rp*-process flow to bypass the waiting point and proceed more quickly through the region. The percentage increase in heavier mass ashes is not as apparent due to the higher absolute abundance of ashes at around mass 60. Since the *rp*-process ashes are pushed down into the neutron star crust under continued accretion, changes in ash composition lead to differences in the thermal evolution of the neutron star crust once accretion has ended [38].

In summary, the high-precision measurement of the mass of ^{56}Cu is reported, allowing the calculation of the ^{55}Ni proton capture energy to a precision of 6.5 keV, a factor of 30 improvement over the AME2012 extrapolated value and a factor of

---

[5]Reprinted figure with permission from Valverde et al. [14]. Copyright 2018 by the American Physical Society.

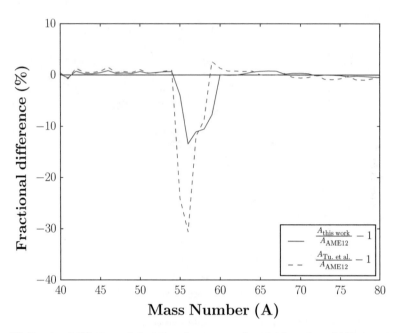

Fig. 4.8 Fractional difference of abundance by mass number of this work (solid blue) compared to that using the masses from AME2012 [12] and the same fractional difference using the mass from Tu et al. [10] (dashed red)

more than 12 improvement over the IMME and CDE calculated values [9, 10] while agreeing with the private communication available in AME2016 [1]. New thermonuclear reaction rates were then calculated using an experimental mass of ^{56}Cu for the first time, and abundances for the *rp*-process around the ^{56}Ni waiting point were determined. These abundances show that the new reaction rate allows the *rp*-process to redirect around this waiting point and proceed to heavier masses more quickly, resulting in an enhancement in higher-mass ashes. The dominant sources of uncertainty are now the unmeasured widths Γ_p and Γ_γ for the ^{55}Ni(p,γ) reaction; the unmeasured higher-lying level scheme of ^{56}Cu; the unmeasured ^{57}Zn mass for the ^{56}Cu(p,γ) reaction and the ^{57}Zn(γ,p) reaction, which hampers this flow from bypassing ^{56}Ni at high temperatures; and the high uncertainty on the β-delayed proton branch of ^{57}Zn (78(17)%, [39]), which directs flow back to ^{56}Ni.

References

1. M. Wang, G. Audi, F. Kondev, W. Huang, S. Naimi, X. Xu, The AME2016 atomic mass evaluation (II). Tables, graphs and references. Chin. Phys. C **41**, 030003 (2017)
2. G.M. Fuller, W.A. Fowler, M.J. Newman, Stellar weak interaction rates for intermediate-mass nuclei. II. a = 21 to a = 60. Astrophys. J. **252**, 715 (1982)

3. A. Kankainen, V.-V. Elomaa, T. Eronen, D. Gorelov, J. Hakala, A. Jokinen, T. Kessler, V.S. Kolhinen, I.D. Moore, S. Rahaman, M. Reponen, J. Rissanen, A. Saastamoinen, C. Weber, J. Äystö, Mass measurements in the vicinity of the doubly magic waiting point ^{56}Ni. Phys. Rev. C **82**, 034311 (2010)

4. R.K. Wallace, S.E. Woosley, Explosive hydrogen burning. Astrophys. J. Suppl. Ser. **45**, 389–420 (1981)

5. H. Schatz, A. Aprahamian, V. Barnard, L. Bildsten, A. Cumming, M. Ouellette, T. Rauscher, F.-K. Thielemann, M. Wiescher, End point of the rp-process on accreting neutron stars. Phys. Rev. Lett. **86**, 3471–3474 (2001)

6. V.-V. Elomaa, G.K. Vorobjev, A. Kankainen, L. Batist, S. Eliseev, T. Eronen, J. Hakala, A. Jokinen, I.D. Moore, Y.N. Novikov, H. Penttilä, A. Popov, S. Rahaman, J. Rissanen, A. Saastamoinen, H. Schatz, D.M. Seliverstov, C. Weber, J. Äystö, Quenching of the SnSbTe cycle in the rp-process. Phys. Rev. Lett. **102**, 252501 (2009)

7. K.E. Rehm, F. Borasi, C.L. Jiang, D. Ackermann, I. Ahmad, B.A. Brown, Brumwell, C.N. Davids, P. Decrock, S.M. Fischer, J. Görres, J. Greene, Hackmann, B. Harss, D. Henderson, W. Henning, R.V.F. Janssens, G. McMichael, V. Nanal, D. Nisius, J. Nolen, R.C. Pardo, M. Paul, P. Reiter, J.P Schiffer, D. Seweryniak, R.E. Segel, M. Wiescher, A.H. Wuosmaa, Study of the ^{56}Ni(dp)^{57}Ni reaction and the astrophysical ^{56}Ni(p, γ)^{57}Cu reaction rate. Phys. Rev. Lett. **80**, 676–679 (1998)

8. C. Langer, F. Montes, A. Aprahamian, D.W. Bardayan, D. Bazin, B.A. Brown, J. Browne, H. Crawford, R.H. Cyburt, C. Domingo-Pardo, A. Gade, S. George, P. Hosmer, L. Keek, A. Kontos, I.-Y. Lee, A. Lemasson, E. Lunderberg, Y. Maeda, M. Matos, Z. Meisel, S. Noji, F.M. Nunes, A. Nystrom, G. Perdikakis, J. Pereira, S.J. Quinn, F. Recchia, H. Schatz, M. Scott, K. Siegl, A. Simon, M. Smith, A. Spyrou, J. Stevens, S.R. Stroberg, D. Weisshaar, J. Wheeler, K. Wimmer, R.G.T. Zegers, Determining the rp-process flow through ^{56}Ni: resonances in ^{57}Cu(p, γ)^{58}Zn identified with GRETINA. Phys. Rev. Lett. **113**, 032502 (2014)

9. W.-J. Ong, C. Langer, F. Montes, A. Aprahamian, D.W. Bardayan, D. Bazin, A. Brown, J. Browne, H. Crawford, R. Cyburt, E.B. Deleeuw, C. Domingo-Pardo, A. Gade, S. George, P. Hosmer, L. Keek, A. Kontos, I.-Y. Lee, A. Lemasson, E. Lunderberg, Y. Maeda, M. Matos, Z. Meisel, S. Noji, F.M. Nunes, A. Nystrom, G. Perdikakis, J. Pereira, S.J. Quinn, F. Recchia, H. Schatz, M. Scott, K. Siegl, A. Simon, M. Smith, A. Spyrou, J. Stevens, S.R. Stroberg, D. Weisshaar, J. Wheeler, K. Wimmer, R.G.T. Zegers, Low- lying level structure of ^{56}Cu and its implications for the rp process. Phys. Rev. C **95**, 055806 (2017)

10. X. Tu, Y. Litvinov, K. Blaum, B. Mei, B. Sun, Y. Sun, M. Wang, H. Xu, Y. Zhang, Indirect mass determination for the neutron-deficient nuclides ^{44}V, ^{48}Mn, ^{52}Co and ^{56}Cu. Nucl. Phys. A **945**, 89–94 (2016)

11. G. Audi, A. Wapstra, C. Thibault, The AME2003 atomic mass evaluation. Nucl. Phys. A **729**, 337–676 (2003)

12. M. Wang, G. Audi, A.H. Wapstra, F.G. Kondev, M. MacCormick, X. Xu, B. Pfeiffer, The AME2012 atomic mass evaluation. Chin. Phys. C **36**, 1603–2014 (2012)

13. H. Schatz, The importance of nuclear masses in the astrophysical rp-process. Int. J. Mass Spectrom. **251**, 293–299 (2006)

14. A.A. Valverde, M. Brodeur, G. Bollen, M. Eibach, K. Gulyuz, A. Hamaker, C. Izzo, W.-J. Ong, D. Puentes, M. Redshaw, R. Ringle, R. Sandler, S. Schwarz, C.S. Sumithrarachchi, J. Surbrook, A.C.C. Villari, I.T. Yandow, High-precision mass measurement of ^{56}Cu and the redirection of the rp-process flow. Phys. Rev. Lett. **120**, 032701 (2018)

15. R. Ringle, G. Bollen, S. Schwarz, Penning trap mass spectrometry of rare isotopes produced via projectile fragmentation at the LEBIT facility. Int. J. Mass Spectrom. **349–350**, 87–93 (2013)

16. D.J. Morrissey, B.M. Sherrill, M. Steiner, A. Stolz, I. Wiedenhoever, Commissioning the A1900 projectile fragment separator. Nucl. Instrum. Methods Phys. Res. Sect. B **204**, 90–96 (2003)

17. K. Cooper, C.S. Sumithrarachchi, D.J. Morrissey, A. Levand, J.A. Rodriguez, G. Savard, S. Schwarz, B. Zabransky, Extraction of thermalized projectile fragments from a large volume gas cell. Nucl. Instrum. Methods Phys. Res. A **763**, 543–546 (2014)

18. S. Schwarz, G. Bollen, R. Ringle, J. Savory, P. Schury, The LEBIT ion cooler and buncher. Nucl. Instrum. Methods Phys. Res. Sect. B **816**, 131–141 (2016)
19. P. Schury, G. Bollen, M. Block, D.J. Morrissey, R. Ringle, A. Prinke, J. Savory, S. Schwarz, T. Sun, Beam purification techniques for low energy rare isotope beams from a gas cell. Hyperfine Interact. **173**, 165–170 (2006)
20. R. Ringle, G. Bollen, A. Prinke, J. Savory, P. Schury, S. Schwarz, T. Sun, The LEBIT 9.4 T Penning trap mass spectrometer. Nucl. Instrum. Methods Phys. Res., Sect. A **604**, 536–547 (2009)
21. K. Blaum, D. Beck, G. Bollen, P. Delahaye, C. Guénaut, F. Herfurth, A. Kellerbauer, H.-J. Kluge, D. Lunney S. Schwarz, L. Schweikhard, C. Yazidjian, Population inversion of nuclear states by a Penning trap mass spectrometer. EPL **67**, 586 (2004)
22. A.A. Kwiatkowski, G. Bollen, M. Redshaw, R. Ringle, S. Schwarz, Isobaric beam purification for high precision Penning trap mass spectrometry of radioactive isotope beams with swift. Int. J. Mass Spectrom. **379** 9–15 (2015)
23. G. Bollen, R.B. Moore, G. Savard, H. Stolzenberg, The accuracy of heavy-ion mass measurements using time of flight-ion cyclotron resonance in a Penning trap. J. Appl. Phys. **68**, 4355–4374 (1990)
24. M. König, G. Bollen, H.-J. Kluge, T. Otto, J. Szerypo, Quadrupole excitation of stored ion motion at the true cyclotron frequency. Int. J. Mass Spectrom. **142**, 95–116 (1995)
25. R.T. Birge, The calculation of errors by the method of least squares. Phys. Rev. **40**, 207–227 (1932)
26. J. Beringer, J.-F. Arguin, R.M. Barnett, K. Copic, O. Dahl, D.E. Groom, C.-J. Lin, J. Lys, H. Murayama, C.G. Wohl, W.-M. Yao et al., Review of particle physics. Phys. Rev. D **86**, 010001 (2012)
27. M. Brodeur, V. Ryjkov, T. Brunner, S. Ettenauer, A. Gallant, V. Simon, M. Smith, A. Lapierre, R. Ringle, P. Delheij, M. Good, D. Lunney, J. Dilling, Verifying the accuracy of the titan Penning-trap mass spectrometer. Int. J. Mass Spectrom. **310**, 20–31 (2012)
28. K. Gulyuz, J. Ariche, G. Bollen, S. Bustabad, M. Eibach, C. Izzo, S.J. Novario, M. Redshaw, R. Ringle, R. Sandler, S. Schwarz, A.A. Valverde, Determination of the direct double β-decay Q value of ^{96}Zr and atomic masses of $^{90-92,94,96}$Zr and $^{92,94-98,100}$Mo. Phys. Rev. C **91**, 055501 (2015)
29. A.A. Valverde, G. Bollen, K. Cooper, M. Eibach, K. Gulyuz, C. Izzo, D.J. Morrissey R. Ringle, R. Sandler, S. Schwarz, C.S. Sumithrarachchi, A.C.C. Villari, Penning trap mass measurement of ^{72}Br. Phys. Rev. C **91**, 037301 (2015)
30. R. Ringle, T. Sun, G. Bollen, D. Davies, M. Facina, J. Huikari, E. Kwan, D.J. Morrissey A. Prinke, J. Savory, P. Schury, S. Schwarz, C.S. Sumithrarachchi, High-precision Penning trap mass measurements of ^{37,38}Ca and their contributions to conserved vector current and isobaric mass multiplet equation. Phys. Rev. C **75**, 055503 (2007)
31. G. Bollen, H.-J. Kluge, M. König, T. Otto, G. Savard, H. Stolzenberg, R.B. Moore, G. Rouleau, G. Audi, I. Collaboration, Resolution of nuclear ground and isometric states by a Penning trap mass spectrometer. Phys. Rev. C. **46**, R2140–R2143 (1992)
32. A.A. Valverde, G. Bollen, M. Brodeur, R.A. Bryce, K. Cooper, M. Eibach, K. Gulyuz, C. Izzo, D.J. Morrissey M. Redshaw, R. Ringle, R. Sandler, S. Schwarz, C.S. Sumithrarachchi, A.C.C. Villari, First direct determination of the superallowed β-decay Q_{EC} value for ^{14}O. Phys. Rev. Lett. **114**, 232502 (2015)
33. M. Honma, T. Otsuka, B.A. Brown, T. Mizusaki, Shell-model description of neutron-rich pf-shell nuclei with a new effective interaction GXPF 1. Eur. Phys. J. A **25** (2005)
34. C. Iliadis, R. Longland, A. Coc, F.X. Timmes, A.E. Champagne, Statistical methods for thermonuclear reaction rates and nucleosynthesis simulations. J. Phys. G **42**, 034007 (2015)
35. J. Fisker, V. Barnard, J. Görres, K. Langanke, G. Marínes-Pinedo, M. Wiescher, Shell model based reaction rates for rp-process nuclei in the mass range $a = 44$–63. At. Data Nucl. Data Tables **79**, 241–292 (2001)
36. T. Rauscher, F.-K. Thielemann, Astrophysical reaction rates from statistical model calculations. At. Data Nucl. Data Tables **75**, 1–351 (2000)

37. R.H. Cyburt, A.M. Amthor, A. Heger, E. Johnson, L. Keek, Z. Meisel, H. Schatz, K. Smith, Dependence of x-ray burst models on nuclear reaction rates. Astrophys. J. **830**, 55 (2016)
38. E.F. Brown, A. Cumming, Mapping crustal heating with the cooling light curves of quasi-persistent transients. Astrophys. J. **698**, 1020 (2009)
39. B. Blank, C. Borcea, G. Canchel, C.E. Demonchy, F. de Oliveira Santos, C. Dossat, J. Giovinazzo, S. Grévy, L. Hay, P. Hellmuth, S. Leblanc, I. Matea, J.L. Pedroza, L. Perrot, J. Pibernat, A. Rebii, L. Serani, J. C. Thomas, Production cross-sections of proton-rich ^{70}Ge fragments and the decay of ^{57}Zn and ^{61}Ge. Eur. Phys. J. A **31**, 267–272 (2007)

Chapter 5
A Cooler-Buncher for the $N = 126$ Factory

5.1 Overview of $N = 126$ Factory

5.1.1 Production

The $N = 126$ factory is a facility under development at the Argonne Tandem Linear Accelerator System (ATLAS) at Argonne National Laboratory, intended to produce nuclei around the $N = 126$ shell closure. The properties of these nuclei, particularly their masses, are critically important for understanding the rapid neutron capture or r-process [1, 2]. Measurements of these masses, however, are currently impossible based on the currently available production techniques for rare isotope beams. In the case of target- or projectile-fragmentation, the relevant targets are unavailable, and relevant beams will have to wait for next-generation facilities like FRIB. Fusion, the other common production techniques for heavy nuclei, is also unable to produce these isotopes. In all these cases, the production cross-sections of the isotopes of interest in the $N = 126$ region are too low to allow for mass measurements [3]. An alternate production method, multi-nucleon transfer (MNT) reactions, makes use of the transfer or multiple nucleons between heavy beams and heavy targets in deep inelastic collisions near the Coulomb barrier [4, 5]. As seen in Fig. 1.7, tests using the EXOGAM high-efficiency germanium array at GANIL show a significant increase in production cross-section for $N = 126$ isotopes using MNTs between a ^{136}Xe beam on a ^{198}Pt target over the projectile fragmentation of a ^{208}Pb beam on a ^{9}Be target at GSI [6–9]. Recent calculations [10] have shown that such a reaction using a 9 MeV/u, 5 pμA ^{136}Xe beam from ATLAS impinged on a stable ^{198}Pt target will produce measurable quantities of the isotopes of interest around the $N = 126$ shell closure (Fig. 5.1[1]).

---

[1]Figure courtesy Kelly [10].

© Springer Nature Switzerland AG 2019
A. A. Valverde, *Precision Measurements to Test the Standard Model and for Explosive Nuclear Astrophysics*, Springer Theses,
https://doi.org/10.1007/978-3-030-30778-3_5

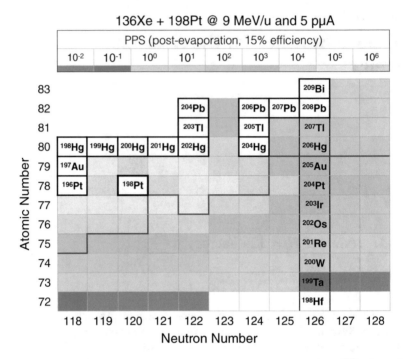

Fig. 5.1 Production (in particles per second) of impinging a 9 MeV/u, 5 pμA ^{136}Xe beam from ATLAS on a stable ^{198}Pt target [11]. The red line indicates the division between measured and unmeasured masses [12]

5.1.2   Design

The $N = 126$ beam factory will have to convert the products of the MNTs, which due to the deeply inelastic nature of the collisions are distributed across wide angles. It will thus be necessary to convert them into a beam, separate out a single isotope, and deliver bunches of this isotope to the experimental station, which in the case of mass measurements will be the Canadian Penning Trap (CPT). This set of requirements is similar to those faced by the Californium Rare Isotope Breeder Upgrade (CARIBU) at ATLAS [13], on which the design of the new facility will be based. A schematic diagram of the $N = 126$ factory can be seen in Fig. 5.2, illustrating its primary elements, which will be further described.

5.1.2.1   Gas Catcher

The $N = 126$ factory begins with the target for MNT reactions, ^{198}Pt and the ^{136}Xe beam from ATLAS which is impinged upon it. The resulting reaction products will be distributed over a wide angular distribution and have a wide

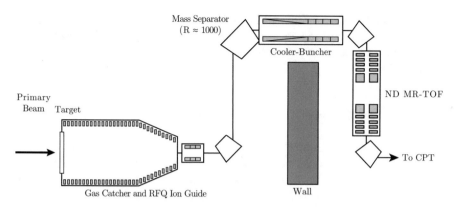

Fig. 5.2 Schematic diagram of the $N = 126$ facility at ATLAS, highlighting the primary elements

distribution of energies; they must be collected and focused into a beam. This will be accomplished using a large-volume helium-filled radiofrequency (RF) gas catcher [14], the downstream side of which will be the platinum target mounting location.

The $N = 126$ gas catcher design will use the linear RF gas catcher [15] design developed at Argonne and is currently in use at CARIBU [16] and for injection into the CPT [17]. As in the CARIBU gas catcher, this design is optimized for reaction products that do not have a preferred direction but instead can fill the entire 2π solid angle after the target. The gas catcher is composed of a cylindrical body filled with $\sim$100 mbar of high-purity helium gas. This gas will slow the reaction products through collisions, and highly charged ions will recombine down to 1^+ or 2^+ charge state, based on second ionization potential of the species in question and the high first- and second-ionization energies of helium. These stopped ions are then guided towards extraction using a combination of RF and DC electric fields and gas flow, entering an RF cone that focuses the ions on the extraction nozzle, from which gas flow will extract them from the stopping volume [16]. On extraction, the ions are guided by a radiofrequency quadrupole (RFQ) ion guide through a region of differential pumping, and then reaccelerated to tens of keV for transport. While a beam dump will be used to stop the primary beam, the rest of the $N = 126$ factory guides the extracted reaction products over a concrete wall designed to protect downstream experimental apparatus from neutrons produced in the MNT reactions (See Fig. 5.2).

5.1.2.2   Dipole Mass Separator Magnet

The next device in the $N = 126$ beamline is a dipole mass separator magnet. This is necessary because the downstream devices have a limited acceptance, while the ions extracted from the gas catcher represent the entire mass range produced by the ^{136}Xe + ^{198}Pt reaction. A 90° dipole bending magnet, in combination with a slit system and an electrostatic quad doublet, will provide separation at $R \sim 1000$, more than sufficient to separate isobars emerging from the gas catcher, but not able to fully resolve isotopes. It will thus significantly reduce the rate of masses incident downstream.

5.1.2.3 Cooler-Buncher

Following the mass separator magnet, the ion beam is injected into an ion beam cooler and buncher. A cooler-buncher is a buffer-gas-filled linear Paul trap designed to convert a high-emittance continuous source into a low-energy, low-emittance ion bunches [18]. Collisions with the buffer gas reduce the transverse emittance and energy spread, providing a damping force and "cooling" the ions, while the RFQ rods of the Paul trap are segmented and a potential is applied such that a weak electric field drags the ions axially to the end of the cooler-buncher [19]. Here, a potential well is created accumulating or "bunching" ions; these bunches are then released downstream by switching the trap potential. Cooler-bunchers are currently in use at many facilities in this role, including the CARIBU [16], ISOLTRAP [18, 20], JYFLTRAP [21], LEBIT [19], SHIPTRAP [22], TITAN [23], and TRIGA-SPEC [24, 25] facilities. The design selected for the $N = 126$ factory is currently in use at the NSCL [26–28].

5.1.2.4 Notre Dame Multi-Reflection Time-of-Flight Mass Spectrometer

The final element of the $N = 126$ factory is the Notre Dame Multi-Reflection Time-of-Flight Mass Spectrometer (MR-TOF) [10, 29]. Its role is to provide isobaric purification of the bunched beam produced from the cooler-buncher for delivery to experiments downstream like the CPT. An MR-TOF does this by the difference in time of flight of the slightly different masses; a pair of electrostatic mirrors folds the flight path of the ions such that the distance traveled by the ions is long but the physical space occupied by the device is small [30]. These devices have typical ion flight times of less than 10 ms, and can yield resolving powers $R > 10^5$, making them a common choice for isobaric separation at many rare isotope beam facilities [31–35], including at CARIBU [36].

The design of the MR-TOF that will be used at the $N = 126$ factory is based on the ISOLDE/CERN MR-TOF [37], with modifications made to the assembly [29]. Each electrostatic mirror is composed of five electrodes and an einzel lens, and the ions are trapped and extracted using the "in-trap lift method" with the central drift tube, where the potential of the drift tube is pulsed from high voltage to ground as the ions pass through it to trap the ions, and from ground to high voltage for extraction [38]. The expected final resolving power of the MR-TOF is $R \sim 10^5$ [29], allowing the separation of isobars for delivery to downstream experiments. The MR-TOF was designed, assembled, and commissioned at the University of Notre Dame, and is the subject of James Kelly's dissertation [10].

5.2 Cooler-Buncher Physics

As previously discussed, an ion beam cooler reduces the emittance or transverse phase space and energy spread of an ion beam; when the longitudinal confinement is used to accumulate and then release bunches of ions, the resulting device is called a cooler-buncher. Cooler-bunchers thus commonly make use of linear radiofrequency quadrupoles for radial confinement, static electrical fields for longitudinal confinement, and buffer gas cooling.

5.2.1 Radiofrequency Quadrupole

An RFQ in its simplest form is composed of four rods placed such that they are the vertices of a square transverse to and centered on the path of the beam. Opposite phases of an oscillating potential are applied to adjacent rods, as illustrated in Fig. 5.3, creating an oscillating potential that results in a time-averaged pseudopotential that under the correct conditions provides radial confinement. Oscillatory electric fields are necessary, as Earnshaw's theorem [39] states that three-dimensional confinement of a collection of charges cannot be provided by electrostatic fields. The quadrupole is the lowest-order multipole that creates a potential minimum, but such a potential yields a saddle point, and so, ion motion is only stable on one axis of the plane, and unstable along the perpendicular axis. By using time-varying inhomogeneous fields, one can create a nonzero time-averaged pseudopotential of parabolic form, confining the ions radially [40]. For heavy ions, these oscillating quadrupole fields commonly have frequencies in the MHz, and hence, these are called radiofrequency quadrupoles or RFQs.

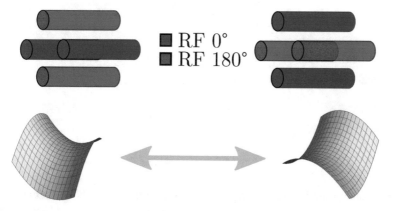

Fig. 5.3 Schematic diagram of RF-switching quadrupole (top) and switching saddle point potential (bottom)

While radial confinement is possible with the application of RF fields, the application of arbitrary RF fields is not sufficient to guarantee this confinement. To determine the conditions of stability, it is necessary to consider the equations of motion. For a quadrupole, the general form of the electrical potential in Cartesian coordinates is [41, 42]

$$\phi = V_0(\alpha x^2 + \beta y^2 + \gamma z^2) \tag{5.1}$$

where V_0 is position-independent and can be time-varying, and α, β, and γ are constants. As this field must obey Laplace's condition $\nabla^2 \phi = 0$, these constants must combine $\alpha + \beta + \gamma = 0$; this can be satisfied in several ways, but for the two-dimensional field, we set $\alpha = 1 = -\beta$ and $\gamma = 0$. If we further define the potential difference between electrode pairs as ϕ_0 and the electrodes as hyperboloids $y^2 - x^2 = \pm r_0^2$ where the minimum distance between electrodes is $2r_0$ (or, equivalently, define the distance from the central axis to an electrode as r_0), this gives an ideal quadrupole potential of

$$\phi = \frac{\phi_0}{2r_0^2}\left(x^2 - y^2\right) \tag{5.2}$$

In the case of an RFQ, the potential ϕ_0 takes the form $\phi_0 = V_{RF}\cos(\omega t) + V_{DC}$, with a constant potential or DC component V_{DC} and a time-varying RF component with magnitude V_{RF} and frequency ω. Combining Newton's second law of motion with the Lorentz force, it can be shown that an ion of charge e and mass m will move with the equations of motion

$$\ddot{x} + \frac{e}{mr_0^2}\left(V_{RF}\cos(\omega t) + V_{DC}\right)x = 0 \tag{5.3}$$

$$\ddot{y} - \frac{e}{mr_0^2}\left(V_{RF}\cos(\omega t) + V_{DC}\right)y = 0 \tag{5.4}$$

We can substitute $a = \frac{4eV_{DC}}{mr_0^2\omega^2}$, $q = \frac{2eV_{RF}}{mr_0^2\omega^2}$, and $\tau = \frac{\omega t}{2}$, and in the form of the canonical Mathieu equations, we have

$$\frac{d^2x}{d\tau^2} + (a + 2q\cos(2\tau))x = 0 \tag{5.5}$$

$$\frac{d^2y}{d\tau^2} - (a + 2q\cos(2\tau))y = 0 \tag{5.6}$$

The Mathieu equation has two types of solutions; these are stable solutions, where the particles oscillate in the $x - y$ plane with limited amplitudes and are confined radially, and unstable solutions, where the amplitude grows exponentially in the x, y, or both directions and the particles are not confined. Which of these two scenarios occurs is determined only by a and q, leading to the shaded stability regions seen in

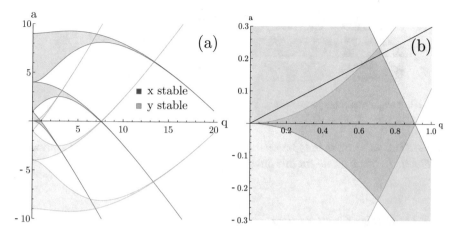

Fig. 5.4 RFQ stability diagrams. (**a**) shows the superposition of the stable regions in both y (blue) and x (gold) directions, and (**b**) shows an expanded view of the stable region near the origin and a black line showing an operating line for a mass filter

the left panel of Fig. 5.4 [42]. RFQs are commonly operated in the largest stability region, located nearest to the origin, expanded in the right panel Fig. 5.4. As $a/q = 2V_{DC}/V_{RF}$, for a constant set of potentials, frequency and mass will only vary the location on the Mathieu stability diagram along a line intersecting with the origin. If the potentials are set such that the operation line lies near the pointed region in the right panel of Fig. 5.4, then a small variation in mass can cause an ion to enter or exit the stability region, creating a mass spectrometer [42]; alternatively, if V_{DC} is zero, which is the common mode of operation for cooler-bunchers, then stability is determined by the region where $q < 0.908$ [43].

5.2.2   Longitudinal Confinement

In a cooler-buncher, the RFQ rods are segmented and static electric fields are applied to provide longitudinal trapping. A simple example of such a trap can be seen in Fig. 5.5, where the RFQ rods are split into three segments with the same RF phases, but different static voltages are applied. If no offset is applied to the first and third sets of electrode, and the middle set has an offset of $-V_{trap}$, a symmetric trapping potential is formed. This potential would take the form [26]

$$V_{trap}(r, z) = \frac{V_{trap}}{z_0^2} \left(z^2 - \frac{r^2}{2} \right) \tag{5.7}$$

with the characteristic distance z_0 determined by the geometry of the trap. While this provides longitudinal trapping, it will also destabilize motion in the transverse

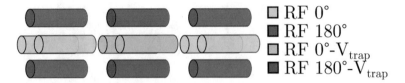

Fig. 5.5 Diagram of a simple symmetric trap created using segmented RFQ rods

plane. The RFQ must thus be operated such that the trapping pseudopotential is sufficient to compensate for this effect.

5.2.3   Buffer Gas Cooling

While an RFQ structure with a longitudinal trapping potential can provide the expected trapping and bunching of ions in a cooler-buncher, it cannot produce the "cooling" or reduction of the transverse phase space or emittance of these ions. This is because of Liouville's theorem, which states that when only conservative forces act on a collection of particles, their emittance is conserved. Thus, to reduce the phase space of this system of ions, a non-conservative force must be introduced. This is done through the introduction of a buffer gas, which adds a non-conservative force in the form of a damping force. Ions in a buffer gas alone would simply follow a random walk; the presence of a trapping potential, however, causes the ions to settle in the potential minimum, and oscillations around that minimum are damped by collisions with the buffer gas, reducing the transverse phase space as desired.

Noble gasses are the usual gas of choice for buffer gas cooling. This is because they have high ionization potentials, which means that it is energetically unfavorable and thus unlikely for a singly charged ion with a lower ionization potential to be neutralized and thus lost. Helium, as the noble gas with the highest ionization potential, is particularly common as a buffer gas. It has the added benefit of being the lightest noble gas, which means that in collision with the $A \sim 200$ ions, the helium should have most of the momentum transferred to them after a collision. While this results in longer cooling times, it also reduces the probability that an ion will scatter out of the potential well.

5.3   Design

The design selected for the $N = 126$ factory cooler-buncher is the same design that is currently in use at the NSCL for the ReA post-accelerator Electron Beam Ion Trap (EBIT) [28] cooler-buncher, which in turn was based on and substantially identical to the design used for the cooler-buncher at the Beam Cooler and Laser

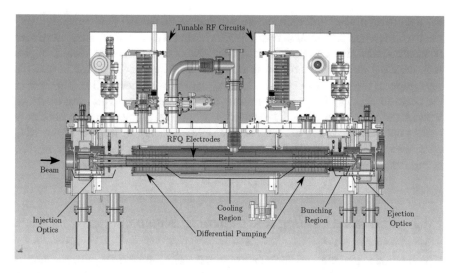

Fig. 5.6 Cross-section view of cooler-buncher design. This view shows the RFQ electrode structure (teal), the PEEK disks that form the differential pumping barriers (off-white), and the injection and ejection electrodes. It also highlights the separate cooling and bunching regions

Spectroscopy (BECOLA) facility at the NSCL [27]. A cross-section of the design can be seen in Fig. 5.6.[2] While the considerations and simulations resulting in the design are discussed at length in Brad Barquest's dissertation [26], an overview of three of the primary novelties of this design follows.

5.3.1 Maximizing Injection Acceptance

As a device designed to operate using rare isotope beams, it is important to maximize the transmission efficiency of the cooler-buncher. An important aspect of this is minimizing losses on injection into the cooler-buncher. This is done by ensuring the emittance of the incident beam fits within the acceptance of the cooler-buncher; as the upstream optics for the $N = 126$ factory were not determined when the cooler-buncher design was determined, as was the case with the design of the BECOLA cooler-buncher, maximizing the injection acceptance is particularly important [26]. The injection optics of a device determine its characteristics; in this design, these optics consist of an immersion lens decelerating the injected beam and hyperboloid and cone electrodes [27]. The hyperboloid ring electrode creates a cylindrically symmetric quadrupole field, which both decelerates and focuses the beam into the RFQ region; the cone electrode is along an equipotential of

---

[2]Reprinted from Valverde et al. [44]. Copyright 2019, with permission from Elsevier.

the hyperboloid ring, and reduces the penetration of the RF field from the RFQ electrodes into the deceleration region [45]. The first segment of the RFQ electrodes themselves was also designed to maximize acceptance. This first section begins with the RFQ electrodes flared away from the beam axis and tapering back to the separation of the other RFQ electrodes over several centimeters [26]. This allows the beam to expand slightly after passing through the hyperboloid ring electrode without colliding with the RFQ electrodes, while still maintaining the necessary radial confinement, which necessitates the tighter spacing of the rest of the RFQ electrodes, as this produces a deeper trapping pseudopotential. A cross-section view of this design and of the assembled injection optics can be seen in Fig. 5.7.[3]

5.3.2   Separated Cooling and Bunching Sections

It is common to operate a cooler-buncher with the same pressure in both the cooling and bunching regions [16, 18, 20, 21, 23, 46]. Higher pressure buffer gas results in shorter cooling times, but higher pressure in the bunching region results in collision-induced "reheating" or increased emittance on ejection. However, many "next-generation" designs make use of differential pumping between the cooling and bunching regions [19, 27]. This allows the cooling region to have a high enough pressure that the cooling time is short enough for experiments with short-lived isotopes to be conducted, while simultaneously the buncher region has a low enough pressure to reduce the reheating of the ions on ejection [26].

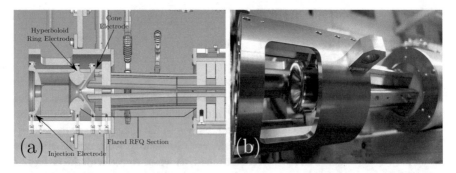

Fig. 5.7 (**a**) Cross-section view of design of injection optics, showing the injection electrode, hyperboloid ring electrode, cone electrode, and flared RFQ section. (**b**) Photograph of assembled injection optics

---

[3]Reprinted from Valverde et al. [44]. Copyright 2019, with permission from Elsevier.

5.3.3 Simplified RFQ Rod Construction

As previously discussed, a longitudinal static potential well is necessary to provide trapping along the axis of the cooler-buncher. Generally, this takes the form of a shallow drag field through the cooling section, providing both trapping and pushing ions forward, and then a sharper, deeper potential well in the bunching section that can be pulsed to eject bunches. The common approach to generating these potentials is to segment the RFQ rods perpendicular to their long axis and add a separate static potential to each segment, or to connect each segment to the next with resistors, and thus gradually decrease the applied potential [16, 18, 20, 21, 24, 47, 48]. While either of these approaches achieve the desired results, they rely on an increased number of internal components, and on a large number of either external (separate static potentials) or internal (resistor-chain) electrical connections, which increases the number of possible failure points and complicates any maintenance. To reduce the number of electrodes, it is instead possible to segment the RFQ electrode diagonally along their long axis, creating two wedges to which different potentials can be applied, creating a uniform drag field with a significant reduction in electrical connections; this is the approach adopted in this cooler-buncher design [26]. An illustration of these two approaches can be seen in Fig. 5.8.

When the RFQ electrodes are segmented, it is necessary to ensure that each RF rod is driven with the same RF signal, independent of the DC potential applied to an individual segment. Commonly, this is achieved through a network of capacitors or transformers isolating the static offset voltage applied to each segment from the RF amplifier. In the design for this cooler-buncher, however, an approach that reduces the complexity of the system has been adopted. A common RF "backbone" electrode runs the length of the RFQ rod, and the various RFQ segments couple capacitively to this electrode; the static offset voltage is applied through leads passing through the RF backbone but separated by ceramic insulators (shown in Fig. 5.9) [26]. Figure 5.10 shows the overall construction of the RFQ electrodes, illustrating both the wedge design and the RF backbone. Figure 5.11[4] illustrates the cut dividing the electrode into wedges as a series of cross-sections moving along the RFQ electrodes.

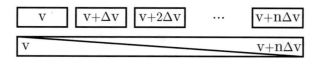

Fig. 5.8 Diagram comparing multiple electrode segments and using diagonally cut electrodes to create a drag potential in a cooler-buncher

---

[4]Reprinted from Valverde et al. [44]. Copyright 2019, with permission from Elsevier.

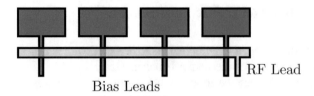

Fig. 5.9 Diagram showing RF backbone coupling scheme. Bias voltage creating the longitudinal trap is applied to the individual segments (blue) through leads passing through the backbone electrode, to which the RF voltage is applied (gold)

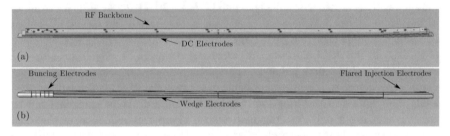

Fig. 5.10 RFQ electrode design, showing (**a**) the RF backbone and capacitively coupled DC electrodes and (**b**) the DC electrodes, showing the segmented bunching section and the wedge-split drag electrodes (highlighted in red)

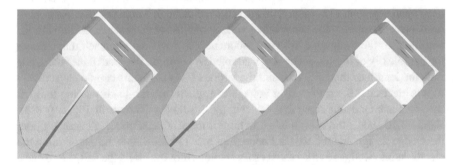

Fig. 5.11 Cross-sections of the RFQ electrode taken perpendicular to the beamline moving from downstream (left) to upstream (right), illustrating the diagonal segmentation creating the wedge-shaped electrodes

5.4   Overview

Figure 5.12 shows an overview of the design of this cooler-buncher along with the various pressure regions. The injection optics, including hyperboloid and cone lenses, decelerate and focus the incident beam, and the flared RFQ structures, which also provide radial confinement, are designed to maximize acceptance. The upstream RFQ structure is split diagonally parallel to the beam axis to simplify the creation of a drag field along the cooling region, where collisions with the buffer

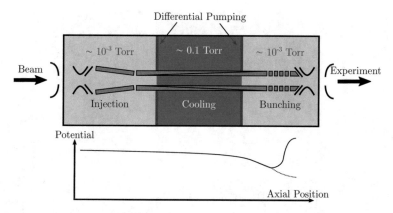

Fig. 5.12 A schematic of the overall design of the cooler-buncher. The top shows the structure of the electrodes, the differential pumping barriers, and the different pressure regions, while the bottom gives a sketch of the DC potential, with the lower (dashed) line showing the potential switched to when the ion bunch is released

gas remove energy from the system. Differential pumping allows the bunching section to operate at a lower pressure, and segmented RFQ rods allow the creation of a potential well to trap ions, and rapid switching on these electrodes enables the release of ion bunches. The RFQ rods use RF backbones to distribute RF capacitively to the RFQ electrodes, reducing the number of connections necessary outside the RFQ.

5.5   Assembly and Commissioning

Assembly of the cooler-buncher followed the design as described in the preceding sections of this chapter. Figure 5.13[5] shows the assembled electrode structure outside of the chamber, comparable to the cross-section that can be seen in Fig. 5.6. Once the assembly of the electrode structure was completed, the wiring of the electrodes and associated RF and DC circuitry could be completed.

Electrical connections for the static and RF potentials of the RFQ electrode structure could then be made. The RF signal is applied through hollow vented tubes connecting to opposite RF backbones in y-shaped arms; the wires for the DC signals are passed through the centers of these tubes. The DC signals are connected to the electrodes through threaded rods passing through the RF backbone that are electrically isolated from it using ceramic spacers. These signals are brought into vacuum through a feedthrough flange, with the individual DC signals being applied to the feedthrough pins, and the RF applied through the flange itself, which connects

---

[5]Reprinted from Valverde et al. [44]. Copyright 2019, with permission from Elsevier.

Fig. 5.13 Photograph of the assembled internal structure of the cooler-buncher before the wiring of the electrodes, including enclosed cooling section, differential pumping barriers for the different pressure regions, and assembly of the electrode structure including flared injection RFQ electrodes and diagonally split cooling region electrodes

to the RF arms but is isolated from the rest of the lid by a ceramic HV break. This assembly can be seen in Fig. 5.14.

The RF is applied to the flange through a tunable resonant LC circuit. The signal from the RF amplifier is sent through an impedance matching transmission-line-transformer-type balun. A balun is a device for connecting a balanced signal, where the conductors carry equal and opposite signals, and an unbalanced line, such as a coaxial cable, where one conductor carries a signal and the other is grounded. In this case, the incoming signal is unbalanced and the outgoing signal is balanced. Furthermore, as this is a transmission-line-transformer-type balun, it provides impedance matching between the low-impedance resonant circuit and the 50 Ω impedance RF signal output by the amplifier, maximizing power transfer and minimizing signal reflection. The opposite phases from the outputs are connected to identical inductances, which are then each connected to the 2.75" flanges after the HV breaks where the RF arms are mounted with a tunable capacitor in parallel. The RFQ and capacitor provide the capacitive load of this resonant LC circuit. As seen in Fig. 5.15, a network analyzer was used to scan over a range of frequencies from 2.5 to 10 MHz, and a resonance was found. The frequency at which this resonance occurs can be changed by varying the capacitance on the tunable capacitor, which can range from 7 to 1000 pF.

Figure 5.16 shows the output of an oscilloscope with inputs from the two flanges with RF applied, where the opposite phases necessary for the creation of quadrupole field for radial trapping can be seen. For testing purposes, an RF amplitude of

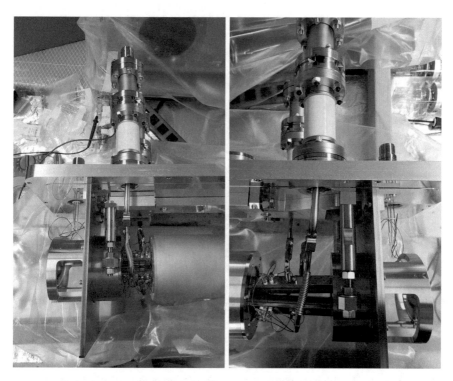

Fig. 5.14 Photograph of wired upstream (right) and downstream (left) electrode structures, showing hollow y-shaped arms applying RF and wires passing through these arms applying DC potentials, as well as the flange and HV break assembly connecting these to air

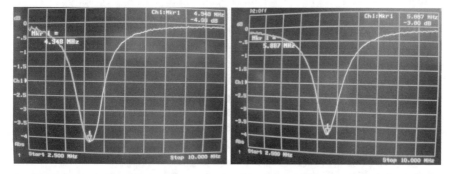

Fig. 5.15 Photograph of network analyzer, showing reflection coefficient in decibels over a range of 2.5–10 MHz for the resonant LC circuit. A resonance can be seen at left at 4.95 MHz, and at right at 5.89 MHz, identifiable by the significant drop in reflected power when a signal at the resonant frequency is transmitted. The movement of the resonance was accomplished by changing the capacitance of the tunable capacitor

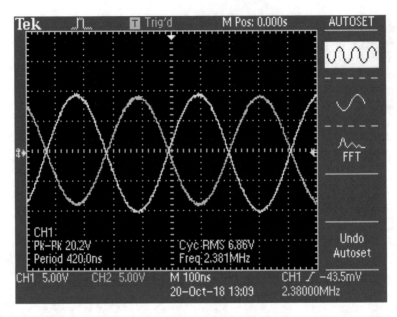

Fig. 5.16 Output of an oscilloscope showing the two opposite phases of RF signal (blue and yellow lines) applied to the two RF flanges of the cooler-buncher

$20V_{pp}$ was used, which resulted in an amplitude mismatch of 8% between these two phases in the test setup. For normal operation, amplitudes of a few hundred volts will be necessary. Tests of the BECOLA cooler-buncher have shown that optimal transmission of $A/q \sim 40$ beam occurs with an amplitude of $100V_{pp}$ and a frequency of 1.2 MHz [26], but normal operation of the EBIT cooler-buncher has demonstrated the transmission of isotopes with a range of masses from 35 to 140 using frequencies of 4–5 MHz and amplitudes of several hundred volts [28].

In summary, a cooler-buncher for the $N = 126$ factory at the ATLAS facility of Argonne National Laboratory has been assembled and tested. The $N = 126$ factory will use multi-nucleon transfer reactions to produce isotopes around the $N = 126$ shell closure that are of great interest for studying the r-process but cannot be produced in sufficient quantities using more traditional production techniques like particle-fragmentation, target-fragmentation, or fission fragments. The MNT reaction products will be collected in a gas cell, extracted, and then separated into isobars using a mass separator magnet. A cooler-buncher is necessary to then take this continuous, fast beam of ions and create bunched, low-energy and emittance ions that can be trapped downstream. This is done using a buffer-gas filled RFQ, where the RFQ provides radial confinement and is segmented to provide for the application of longitudinal confinement. The design used for the $N = 126$ factory is the same as is currently in use at the NSCL before their EBIT, and is based on the design used for the BECOLA cooler-buncher. This design features several improvements over previous designs, including separated, different-pressure cooling and bunching regions, flared injection RFQ electrodes for increased acceptance, and a simplified RFQ electrode design.

References

1. M. Mumpower, R. Surman, G. McLaughlin, A. Aprahamian, The impact of individual nuclear properties on r-process nucleosynthesis. Prog. Part. Nucl. Phys. **86**, 86–126 (2016)
2. M. Arnould, S. Goriely, K. Takahashi, The r-process of stellar nucleosynthesis: astrophysics and nuclear physics achievements and mysteries. Phys. Rep. **450**, 97–213 (2007)
3. Y.X. Watanabe, Y.H. Kim, S.C. Jeong, Y. Hirayama, N. Imai, H. Ishiyama, H.S. Jung, H. Miyatake, S. Choi, J.S. Song, E. Clement, G. de France, A. Navin, M. Rejmund, C. Schmitt, G. Pollarolo, L. Corradi, E. Fioretto, D. Montanari, M. Niikura, D. Suzuki, H. Nishibata, J. Takatsu, Pathway for the production of neutron-rich isotopes around the $N = 126$ shell closure. Phys. Rev. Lett. **115**, 172503 (2015)
4. C.H. Dasso, G. Pollarolo, A. Winther, Systematics of isotope production with radioactive beams. Phys. Rev. Lett. **73**, 1907–1910 (1994)
5. V. Zagrebaev, W. Greiner, Production of new heavy isotopes in low-energy multinucleon transfer reactions. Phys. Rev. Lett. **101**, 122701 (2008)
6. Y. Hirayama, H. Miyatake, Y.X. Watanabe, N. Imai, H. Ishiyama, S.C. Jeong, H.S. Jung, M. Oyaizu, M. Mukai, S. Kimura, T. Sonoda, M. Wada, Y.H. Kim, M. Huyse, Yu. Kudryavtsev, P. Van Duppen, Beta-decay spectroscopy of r-process nuclei around $n = 126$. EPJ. Web Conf. **109**, 08001 (2016)
7. Y. Hirayama, Y.X. Watanabe, N. Imai, H. Ishiyama, S.C. Jeong, H. Miyatake, M. Oyaizu, M. Mukai, S. Kimura, Y.H. Kim, T. Sonoda, M. Wada, M. Huyse, Y. Kudryavtsev, P.V. Duppen, Beta-decay spectroscopy of r-process nuclei with $N = 126$ at KEK isotope separation system, in *Proceedings of the Conference on Advances in Radioactive Isotope Science (ARIS 2014)*
8. T. Kurtukian-Nieto, J. Benlliure, K.-H. Schmidt, L. Audouin, F. Becker, B. Blank, E. Casarejos, F. Farget, M. Fernández-Ordóñez, J. Giovinazzo, D. Henzlova, B. Jurado, J. Pereira, O. Yordanov, Production cross sections of heavy neutron-rich nuclei approaching the nucleosynthesis r-process path around $A = 195$. Phys. Rev. C **89**, 024616 (2014)
9. Y. Watanabe, Y. Hirayama, N. Imai, H. Ishiyama, S. Jeong, H. Miyatake, E. Clement, G. de France, A. Navin, M. Rejmund, C. Schmitt, G. Pollarolo, L. Corradi, E. Fioretto, D. Montanari, S. Choi, Y. Kim, J. Song, M. Niikura, D. Suzuki, H. Nishibata, J. Takatsu, Study of collisions of ^{136}Xe $+$ ^{198}Pt for the KEK isotope separator. Nucl. Instrum. Methods Phys. Res. Sect. B **317**, 752–755 (2013). XVIth International Conference on Electromagnetic Isotope Separators and Techniques Related to their Applications, December 2–7, 2012 at Matsue, Japan
10. J.M. Kelly, Precision measurements and development of a spectrometer to determine atomic masses of high impact for the astrophysical r-process. Ph.D. thesis, University of Notre Dame, 2019
11. A. Winther, Grazing reactions in collisions between heavy nuclei. Nucl. Phys. A **572**, 191–235 (1994)
12. M. Wang, G. Audi, F. Kondev, W. Huang, S. Naimi, X. Xu, The AME2016 atomic mass evaluation (II). Tables, graphs and references. Chin. Phys. C **41**, 030003 (2017)
13. G. Savard, R.C. Pardo, S. Baker, C.N. Davids, A. Levand, D. Peterson, D.G. Phillips, T. Sun, R. Vondrasek, B.J. Zabransky, G.P. Zinkann, CARIBU: a new facility for the study of neutron-rich isotopes. Hyperfine Interact. **199**, 301–309 (2011)
14. G. Savard, Large radio-frequency gas catchers and the production of radioactive nuclear beams. J. Phys. Conf. Ser. **312**, 052004 (2011)

15. G. Savard, J. Clark, C. Boudreau, F. Buchinger, J. Crawford, H. Geissel, J. Greene, S. Gulick, A. Heinz, J. Lee, A. Levand, M. Maier, G. Münzenberg, C. Scheidenberger, D. Seweryniak, K. Sharma, G. Sprouse J. Vaz, J. Wang, B. Zabransky, Z. Zhou, Development and operation of gas catchers to thermalize fusion–evaporation and fragmentation products. Nucl. Instrum. Methods Phys. Res. Sect. B **204**, 582–586 (2003). 14th International Conference on Electromagnetic Isotope Separators and Techniques Related to their Applications

16. G. Savard, A. Levand, B. Zabransky, The CARIBU gas catcher. Nucl. Instrum. Methods Phys. Res. Sect. B **376**, 246–250 (2016). Proceedings of the XVIIth International Conference on Electromagnetic Isotope Separators and Related Topics (EMIS2015), Grand Rapids, MI, 11–15 May 2015

17. G. Savard, R.C. Barber, C. Boudreau, F. Buchinger, J. Caggiano, J. Clark, J.E. Crawford, H. Fukutani, S. Gulick, J.C. Hardy, A. Heinz, J.K.P. Lee, R.B. Moore, K.S. Sharma, J. Schwartz, D. Seweryniak, G.D. Sprouse, J. Vaz, The Canadian Penning trap spectrometer at Argonne. Hyperfine Interact. **132**, 221–228 (2001)

18. F. Herfurth, J. Dilling, A. Kellerbauer, G. Bollen, S. Henry, H.-J. Kluge, E. Lamour, D. Lunney, R. Moore, C. Scheidenberger, S. Schwarz, G. Sikler, J. Szerypo, A linear radiofrequency ion trap for accumulation, bunching, and emittance improvement of radioactive ion beams. Nucl. Instrum. Methods Phys. Res. Sect. B **469**, 254–275 (2001)

19. S. Schwarz, G. Bollen, R. Ringle, J. Savory, P. Schury, The LEBIT ion cooler and buncher. Nucl. Instrum. Methods Phys. Res. Sect. B **816**, 131–141 (2016)

20. M. Smith, L. Blomeley, P Delheij, J. Dilling, First tests of the titan digital RFQ beam cooler and buncher, in *TCP 2006*, ed. by J. Dilling, M. Comyn, J. Thompson, G. Gwinner (Springer, Berlin, 2007), pp. 327–336

21. A. Nieminen, J. Huikari, A. Jokinen, J. Äystö, P. Campbell, E. Cochrane, Beam cooler for low-energy radioactive ions. Nucl. Instrum. Methods Phys. Res. Sect. B **469**, 244–253 (2001)

22. G. Marx, D. Ackermann, J. Dilling, F. Hessberger, S. Hoffmann, H.-J. Kluge, R. Mann, G. Münzenberg, Z. Qamhieh, W. Quint, D. Rodriguez, M. Schädel, J. Schönfelder, G. Sikler, C. Toader, C. Weber, O. Engels, D. Habs, P. Thirolf, H. Backe, A. Dretzke, W. Lauth, W. Ludolphs, M. Sewtz, Status of the SHIPTRAP project: a capture and storage facility for heavy radionuclides from SHIP. Hyperfine Interact. **132**, 459–464 (2001)

23. T. Brunner, M. Smith, M. Brodeur, S. Ettenauer, A. Gallant, V. Simon, A. Chaudhuri, A. Lapierre, E. Mané, R. Ringle, M. Simon, J. Vaz, P Delheij, Good, M. Pearson, J. Dilling, Titan's digital RFQ ion beam cooler and buncher, operation and performance. Nucl. Instrum. Methods Phys. Res., Sect. A **676**, 32–43 (2012)

24. D. Lunney, C. Bachelet, C. Guénaut, S. Henry, M. Sewtz, COLETTE: a linear Paul-trap beam cooler for the on-line mass spectrometer mistral. Nucl. Instrum. Methods Phys. Res., Sect. A **598**, 379–387 (2009)

25. T. Beyer, K. Blaum, M. Block, C.E. Düllmann, K. Eberhardt, M. Eibach, N. Frömmgen, C. Geppert, C. Gorges, J. Grund, M. Hammen, S. Kaufmann, A. Krieger, S. Nagy, W. Nörterhäuser, D. Renisch, C. Smorra, E. Will, An RFQ cooler and buncher for the TRIGA-spec experiment. Appl. Phys. B **114**, 129–136 (2014)

26. B.R. Barquest, An advanced ion guide for beam cooling and bunching for collinear laser spectroscopy of rare isotopes. Ph.D. thesis, Michigan State University, 2014

27. B. Barquest, G. Bollen, P. Mantica, K. Minamisono, R. Ringle, S. Schwarz, C. Sumithrarachchi, RFQ beam cooler and buncher for collinear laser spectroscopy of rare isotopes. Nucl. Instrum. Methods Phys. Res. Sect. A **866**, 18–28 (2017).

28. A. Lapierre, G. Bollen, D. Crisp, S.W. Krause, L.E. Linhardt, K. Lund, Nash, R. Rencsok, R. Ringle, S. Schwarz, M. Steiner, C. Sumithrarachchi, T. Summers, A.C.C. Villari, S.J. Williams, Q. Zhao, First two operational years of the electron-beam ion trap charge breeder at the national superconducting cyclotron laboratory. Phys. Rev. Accel. Beams **21**, 053401 (2018)

29. B. Schultz, J. Kelly, C. Nicoloff, J. Long, S. Ryan, M. Brodeur, Construction and simulation of a multi-reflection time-of-flight mass spectrometer at the University of Notre Dame. Nucl.

Inst. Methods Phys. Res. B **376**, 251–255 (2016). Proceedings of the XVIIth International Conference on Electromagnetic Isotope Separators and Related Topics (EMIS2015), Grand Rapids, MI, 11–15 May 2015

30. H. Wollnik, M. Przewloka, Time-of-flight mass spectrometers with multiply reflected ion trajectories. Int. J. Mass Spectrom. Ion Process. **96**, 267–274 (1990)
31. W.R. Plaß, T. Dickel, U. Czok, H. Geissel, M. Petrick, K. Reinheimer, C. Scheidenberger, M.I. Yavor, Isobar separation by time-of-flight mass spectrometry for low-energy radioactive ion beam facilities. Nucl. Instrum. Methods Phys. Res. Sect. B **266**, 4560–4564 (2008). Proceedings of the XVth International Conference on Electromagnetic Isotope Separators and Techniques Related to their Applications
32. A. Piechaczek, V. Shchepunov, H. Carter, J. Batchelder, E. Zganjar, S. Liddick, H. Wollnik, Y. Hu, B. Griffith, Development of a high resolution isobar separator for study of exotic decays. Nucl. Instrum. Methods Phys. Res. Sect. B **266**, 4510–4514 (2008). Proceedings of the XVth International Conference on Electromagnetic Isotope Separators and Techniques Related to their Applications
33. R. Wolf, D. Beck, K. Blaum, C. Böhm, C. Borgmann, M. Breitenfeldt, F. Herfurth, A. Herlert, M. Kowalska, S. Kreim, D. Lunney S. Naimi, D. Neidherr, M. Rosenbusch, L. Schweikhard, J. Stanja, F. Wienholtz, K. Zuber, Online separation of short-lived nuclei by a multi-reflection time-of-flight device. Nucl. Instrum. Methods Phys. Res. Sect. A **686**, 82–90 (2012)
34. W.R. Plaß, T. Dickel, C. Scheidenberger, Multiple-reflection time-of-flight mass spectrometry. Int. J. Mass Spectrom. **349–350**, 134–144 (2013). 100 years of Mass Spectrometry
35. M. Rosenbusch, P. Ascher, D. Atanasov, C. Barbieri, D. Beck, K. Blaum, C. Borgmann, M. Breitenfeldt, R.B. Cakirli, A. Cipollone, S. George, F. Herfurth, M. Kowalska, S. Kreim, D. Lunney, V. Manea, P Navrátil, D. Neidherr, L. Schweikhard, V. Somà, J. Stanja, F. Wienholtz, R.N. Wolf, K. Zuber, Probing the $N = 32$ shell closure below the magic proton number $Z = 20$: mass measurements of the exotic isotopes ^{52,53}K. Phys. Rev. Lett. **114**, 202501 (2015)
36. T.Y. Hirsh, N. Paul, M. Burkey, A. Aprahamian, F. Buchinger, S. Caldwell, J.A. Clark, A.F. Levand, L.L. Ying, S.T. Marley, G.E. Morgan, A. Nystrom, R. Orford, A.P Galván, J. Rohrer, G. Savard, K.S. Sharma, K. Siegl, First operation and mass separation with the CARIBU MR-TOF. Nucl. Instrum. Methods Phys. Res. Sect. B **376**, 229–232 (2016). Proceedings of the XVIIth International Conference on Electromagnetic Isotope Separators and Related Topics (EMIS2015), Grand Rapids, MI, USA, 11–15 May 2015
37. R.N. Wolf, M. Eritt, G. Marx, L. Schweikhard, A multi-reflection time-of-flight mass separator for isobaric purification of radioactive ion beams. Hyperfine Interact. **199**, 115–122 (2011)
38. R.N. Wolf, G. Marx, M. Rosenbusch, L. Schweikhard, Static-mirror ion capture and time focusing for electrostatic ion-beam traps and multi-reflection time-of-flight mass analyzers by use of an in-trap potential lift. Int. J. Mass Spectrom. **313**, 8–14 (2012)
39. S. Earnshaw, On the nature of the molecular forces which regulate the constitution of the luminiferous ether. Trans. Camb. Philos. Soc. **7**, 97 (1848).
40. H. Dehmelt, Radiofrequency spectroscopy of stored ions I: storage. Part II: spectroscopy is now scheduled to appear in volume v of this series, in *Advances in Atomic and Molecular Physics*, ed. by D. Bates, I. Estermann, vol. 3 (Academic, London, 1968), pp. 53–72
41. P.H. Dawson, *Quadrupole Mass Spectrometry and Its Applications* (Elsevier, Amsterdam, 1976)
42. W. Paul, Electromagnetic traps for charged and neutral particles. Rev. Mod. Phys. **62**, 531–540 (1990)
43. G. Bollen, Traps for rare isotopes, in *The Euroschool Lectures on Physics with Exotic Beams*, vol. 1, ed. by J.S. Al-Khalili, E. Roeckl (Springer, Berlin, 2004)
44. A.A. Valverde, M. Brodeur, J. Clark, D. Lascar, G. Savard, A cooler-buncher for the $N = 126$ factory at Argonne National Laboratory. Nucl. Instrum. Methods Phys. Res. Sect. B (2019)
45. O. Gianfrancesco, F. Duval, G. Ban, R. Moore, D. Lunney, A radiofrequency quadrupole cooler for high-intensity beams. Nucl. Instrum. Methods Phys. Res. Sect. B **266**, 4483–4487 (2008). Proceedings of the XVth International Conference on Electromagnetic Isotope Separators and Techniques Related to their Applications

46. I.P Aliseda, T. Fritioff, T. Giles, A. Jokinen, M. Lindroos, F. Wenander, Design of a second
    generation RFQ ion cooler and buncher (RFQCB) for ISOLDE. Nucl. Phys. A **746**, 647–
    650 (2004). Proceedings of the Sixth International Conference on Radioactive Nuclear Beams
    (RNB6)
47. G. Sikler, D. Ackermann, F. Attallah, D. Beck, J. Dilling, S. Elisseev, H. Geissel, D. Habs,
    S. Heinz, F. Herfurth, F. Heßberger, S. Hofmann, H.-J. Kluge, C. Kozhuharov, G. Marx, M.
    Mukherjee, J. Neumayr, W. Plaß, W. Quint, S. Rahaman, D. Rodríguez, C. Scheidenberger,
    M. Tarisien, P. Thirolf, V. Varentsov, C. Weber, Z. Zhou, First on-line test of SHIPTRAP. Nucl.
    Instrum. Methods Phys. Res. Sect. B **204**, 482–486 (2003). 14th International Conference on
    Electromagnetic Isotope Separators and Techniques Related to their Applications
48. G. Ban, G. Darius, D. Durand, P. Delahaye, E. Liénard, F. Mauger, O. Naviliat-Cuncic, J.
    Szerypo, First tests of a linear radiofrequency quadrupole for the cooling and bunching of
    radioactive light ions. Hyperfine Interact. **146**, 259–263 (2003)

Chapter 6
Summary and Outlook

Precision measurements in nuclear physics is an active field of study, and is a critical avenue of research for a wide variety of areas. These include the study of nuclear structure, the study of astrophysical reaction rates and nucleosynthesis pathways, the testing of mass models, and the testing of the Standard Model and the search for physics beyond it [1]. This dissertation presents my work on three specific applications of precision measurements.

First presented was a new, high-precision ^{11}C half-life measurement that I conducted at the University of Notre Dame's Nuclear Science Laboratory using the *TwinSol* facility [2]. The new value, $t_{1/2} = 1220.27(26)$ s, is consistent with previous values, but offers a factor of greater than five improvement over the previous most precise measurement; the newly calculated world average, $t_{1/2}^{world} = 1220.41(32)$ s, also shows a fivefold improvement. This, in combination with other experimental and theoretical parameters, allowed me to calculate a $\mathcal{F}t^{mirror}$ value that is the most precise of all superallowed $T = 1/2$ mixed mirror transitions, and comparable to the $\mathcal{F}t^{0^+ \to 0^+}$ values that currently provide the most precise determination of V_{ud}, and thus the most stringent test of CKM matrix unitarity and the electroweak sector of the Standard Model.

This result provides a clear motivation to improve the precision on the calculated $\delta_{NS}^V - \delta_C^V$ correction, which is now the largest source of uncertainty in the $\mathcal{F}t^{mirror}$ value, and for a measurement of the Fermi-to-Gamow-Teller mixing ratio ρ, which would allow for the determination of V_{ud} from this $\mathcal{F}t^{mirror}$ value. Currently, the development of a Paul trap for the measurement of the beta-neutrino angular correlation parameter $a_{\beta\nu}$, St. Benedict (the Superallowed Transition Beta-Neutrino-Decay Ion Coincidence Trap), is underway at the University of Notre Dame, which would allow for the determination of ρ. Also of interest for the determination of V_{ud} and CKM unitarity would be the high-precision half-life measurement of ^{29}P, one of the five superallowed mixed mirror decay isotopes for which ρ is known and where the half-life uncertainty is dominant in the uncertainty on V_{ud}. This measurement

© Springer Nature Switzerland AG 2019
A. A. Valverde, *Precision Measurements to Test the Standard Model and for Explosive Nuclear Astrophysics*, Springer Theses,
https://doi.org/10.1007/978-3-030-30778-3_6

is also planned for the Notre Dame β Counting Station, pending developments that will allow for the clear identification of radioactive contaminants produced with the ^{29}P beam.

The second section of this dissertation presented the high-precision mass measurement of ^{56}Cu that I conducted using the LEBIT 9.4 T Penning trap mass spectrometer at the National Superconducting Cyclotron Laboratory at Michigan State University [3]. This new measurement, ME $= -38,626.7(7.1)$ keV, is the first published experimental mass for this isotope, and resolves a several hundred keV discrepancy between previous extrapolated and calculated masses. This was of particular interest because the mass of ^{56}Cu is important for determining the ^{55}Ni(p,γ) and ^{56}Cu(p,γ) forward and reverse reaction rates, which in turn govern the flow of the rp-process through the ^{55}Ni(p,γ)^{56}Cu(p,γ)^{57}Zn(β^{+})^{57}Cu bypass of the ^{56}Ni waiting point. This new mass was used to calculate reaction rates, and then a precision network calculation was run, which demonstrated that the rp-process does partially redirect around the ^{56}Ni waiting point.

To further refine the pathway of the rp-process around the ^{56}Ni waiting point, the new mass measurement and recent low-lying level scheme of ^{56}Cu by Ong et al. [4] should be complemented by the measurement of the higher-lying level scheme of ^{56}Cu and the widths Γ_p and Γ_γ for the ^{55}Ni(p,γ) reaction, the measurement of the currently unmeasured mass of ^{57}Zn to determine the ^{56}Cu(p,γ) and the ^{57}Zn(γ,p) reactions, and a new measurement of the β-delayed proton branch of ^{57}Zn, which directs flow back towards ^{56}Ni. Recent efforts by Schatz and Ong [5] have run X-ray burst models while varying input masses by 3σ. This showed three masses (^{27}P, ^{61}Ga, and ^{65}As) that have a significant impact on the light curve and ash composition of a typical H/He X-ray burst, and an additional three (^{80}Zr, ^{81}Zr, and ^{82}Nb) that have a significant impact on only the ash composition. A Penning trap mass measurement of ^{27}P has been approved at the NSCL as experiment 18002. Furthermore, the determination of the ^{23}Al(p,γ)^{24}Si reaction rate is also of interest for determining the rp-process light curve; after a recent effort to reduce the uncertainty on this reaction rate using GRETINA and LENDA at the NSCL [6], the leading source of uncertainty is the Q value, which is dominated by the mass uncertainty on ^{24}Si. Experiment 18005 was approved at the NSCL as a Penning trap mass measurement of this isotope to reduce this uncertainty.

The third and final section of this dissertation provided an overview of the $N = 126$ factory, a new facility under construction at Argonne National Laboratory's ATLAS accelerator to produce isotopes around the $N = 126$ shell closure to allow for measurements of interest in constraining the astrophysical r-process pathway. It focused on my assembly and preliminary commissioning of the RFQ cooler-buncher, a key component of the facility that will take the high-emittance continuous beam extracted from the gas catcher and convert it into low-emittance discrete ion bunches, allowing for trapping in downstream components. The general design principles of RFQs were discussed, as well as the specific design used in this device, which was also used at the NSCL's EBIT [7].

With the completion of the cooler-buncher, the remaining major elements of the $N = 126$ factory are the gas catcher and the multi-reflection time-of-flight mass spectrometer (MR-TOF). Assembly of the MR-TOF at the University of Notre Dame is complete, with the completion of commissioning envisioned for the first half of 2019 [8], while assembly of the gas catcher is nearing completion, with commissioning expected to begin early 2019. The design of the beamline is ongoing, and completion of the $N = 126$ beam factory is planned for early 2020, with a mass measurement campaign using the Canadian Penning Trap to begin shortly thereafter.

References

1. K. Blaum, High-accuracy mass spectrometry with stored ions. Phys. Rep. **425**, 1–78 (2006)
2. A.A. Valverde, M. Brodeur, T. Ahn, J. Allen, D.W. Bardayan, F.D. Becchetti, D. Blankstein, G. Brown, D.P. Burdette, B. Frentz, G. Gilardy, M.R. Hall, S. King, J.J. Kolata, J. Long, K.T. Macon, A. Nelson, P.D. O'Malley, M. Skulski, S.Y. Strauss, B. Vande Kolk, Precision half-life measurement of ^{11}C: the most precise mirror transition Ft value. Phys. Rev. C **97**, 035503 (2018)
3. A.A. Valverde, M. Brodeur, G. Bollen, M. Eibach, K. Gulyuz, A. Hamaker, C. Izzo, W.-J. Ong, D. Puentes, M. Redshaw, R. Ringle, R. Sandler, S. Schwarz, C.S. Sumithrarachchi, J. Surbrook, A.C.C. Villari, I.T. Yandow, High-precision mass measurement of ^{56}Cu and the redirection of the rp-process flow. Phys. Rev. Lett. **120**, 032701 (2018)
4. W.-J. Ong, C. Langer, F. Montes, A. Aprahamian, D.W. Bardayan, D. Bazin, B.A. Brown, J. Browne, H. Crawford, R. Cyburt, E.B. Deleeuw, C. Domingo-Pardo, A. Gade, S. George, P. Hosmer, L. Keek, A. Kontos, I.-Y. Lee, A. Lemasson, E. Lunderberg, Y. Maeda, M. Matos, Z. Meisel, S. Noji, F.M. Nunes, G. Perdikakis, J. Pereira, S.J. Quinn, F. Recchia, H. Schatz, M. Scott, K. Siegl, A. Simon, M. Smith, A. Spyrou, J. Stevens, S.R. Stroberg, D. Weisshaar, J. Wheeler, K. Wimmer, R.G.T. Zegers, Low-lying level structure of ^{56}Cu and its implications for the rp process. Phys. Rev. C **95**, 055806 (2017)
5. H. Schatz, W.-J. Ong, Dependence of x-ray burst models on nuclear masses. Astrophys. J. **844**, 139 (2017)
6. C. Wolf, C. Langer, F. Montes, J. Pereira, S. Ahn, S. Ayoub, D. Bazin, P Bender, A. Brown, J. Browne, H. Crawford, E. Deleeuw, B. Elman, S. Fiebiger, A. Gade, P Gastis, S. Lipschutz, B. Longfellow, F. Nunes, W.-J. Ong, T. Poxon-Pearson, G. Perdikakis, R. Reifarth, H. Schatz, K. Schmidt, J. Schmitt, C. Sullivan, R. Titus, D. Weisshaar, P. Woods, J.C. Zamora, G.T. Zegers, Constraining the rp-process by measuring ^{23}Al(d,n)^{24}Si with GRETINA and LENDA at NSCL. EPJ Web Conf. **165**, 01055 (2017)
7. A. Lapierre, G. Bollen, D. Crisp, S.W. Krause, L.E. Linhardt, K. Lund, Nash, R. Rencsok, R. Ringle, S. Schwarz, M. Steiner, C. Sumithrarachchi, T. Summers, A.C.C. Villari, S.J. Williams, Q. Zhao, First two operational years of the electron-beam ion trap charge breeder at the national superconducting cyclotron laboratory. Phys. Rev. Accel. Beams **21**, 053401 (2018)
8. J.M. Kelly, Precision measurements and development of a spectrometer to determine atomic masses of high impact for the astrophysical r-process. Ph.D. thesis, University of Notre Dame, 2019

Curriculum Vitae

Physics Division
Building 203
Argonne National Laboratory
Lemont, IL 60439
✉ *avalverde@anl.gov*

Adrián A. Valverde

━━━━━━━ Education

2016–2018 **Ph.D., Physics**, *University of Notre Dame, Notre Dame, IN.*

Dissertation: *Precision Measurements to Test the Standard Model and for Explosive Nuclear Astrophysics*

2017 **M.S., Physics**, *University of Notre Dame, Notre Dame, IN.*

2012–2015 **M.S., Physics**, *Michigan State University, East Lansing, MI.*

2008–2012 **B.S., Physics**, *Michigan State University, East Lansing, MI.*

2008–2012 **B.S., Mathematics**, *Michigan State University, East Lansing, MI.*

━━━━━━━ Appointments

2019–Present **Resident Associate**, *Low Energy Physics Group*, Argonne National Laboratory, Lemont, IL.

2019–Present **Postdoctoral Researcher**, *Canadian Penning Trap Collaboration*, University of Manitoba, Winnipeg, MB, Canada.

2017–2018 **Visiting Graduate Student**, *Low Energy Physics Group*, Argonne National Laboratory, Lemont, IL.

2016–2018 **Graduate Research Assistant**, *Brodeur Group*, University of Notre Dame, Notre Dame, IN.

2012–2016 **Graduate Research Assistant**, *LEBIT Group*, National Superconducting Cyclotron Laboratory, East Lansing, MI.

2008–2012 **Undergraduate Research Assistant**, *LEBIT Group*, National Superconducting Cyclotron Laboratory, East Lansing, MI.

© Springer Nature Switzerland AG 2019

A. A. Valverde, *Precision Measurements to Test the Standard Model and for Explosive Nuclear Astrophysics*, Springer Theses, https://doi.org/10.1007/978-3-030-30778-3

████████ Teaching Experience

Spring 2017 **PHYS31220: Physics II Lab**, *Lead Teaching Assistant*, University of Notre Dame, Notre Dame, IN.

Fall 2016 **PHYS31220: Physics II Lab**, *Teaching Assistant*, University of Notre Dame, Notre Dame, IN.

Spring 2016 **ISP 209L: Mysteries of the Physical World Lab**, *Instructor*, Michigan State University, East Lansing, MI.

Fall 2015 **PHY 191: Physics Lab for Scientists I**, *Teaching Assistant*, Michigan State University, East Lansing, MI.

████████ Awards and Fellowships

2012 **Rasmussen Fellowship**, *Michigan State University*, East Lansing, MI.

2012 **College of Natural Sciences Fellowship**, *MSU College of Natural Sciences*, East Lansing, MI.

2012 **Carl L. Foiles Endowment Award**, *MSU Department of Physics & Astronomy*, East Lansing, MI.

2011 **Lawrence W. Hantel Endowed Fellowship**, *MSU Department of Physics & Astronomy*, East Lansing, MI.

2008 **University Distinguished Scholarship**, *Michigan State University*, East Lansing, MI.

████████ Approved Experiments

2019 **Commissioning the Cooler-Buncher and MR-TOF Mass Spectrometer for the $N = 126$ Beam Factory**, *A. A. Valverde*, ATLAS PAC 2019, Pr. # 1797.
8 d. approved

2018 **High precision mass measurement of 27P for the astrophysical rp process**, *A. Valverde*, NSCL PAC 42, Ex. #18002.
58 h. approved

████████ Refereed Publications

G. Savard, M. Brodeur, J. A. Clark, R. A. Knaack, and **A. A. Valverde**, "The N=126 factory: A new facility to produce very heavy neutron-rich isotopes", Nucl. Instrum. Methods B, Proceedings of the XVIIIth International Conference on Electromagnetic Isotope Separators and Related Topics (EMIS2018), Geneva, GE, CH., 16–21 September 2018 (2019).

A. A. Valverde, M. Brodeur, J. Clark, D. Lascar, and G. Savard, "A cooler-buncher for the N=126 factory at Argonne National Laboratory", Nucl. Instrum. Methods B, Proceedings of the XVIIIth International Conference on Electromagnetic Isotope Separators and Related Topics (EMIS2018), Geneva, GE, CH., 16–21 September 2018 (2019).

D. Burdette, M. Brodeur, P. O'Malley, and **A. A. Valverde**, "Development of the St. Benedict Paul Trap at the Nuclear Science Laboratory", Hyperfine Interact. **240**, Proceedings of the 7th International Conference on Trapped Charged Particles and Fundamental Physics (TCP

2018), Traverse City, Michigan, USA, 30 September–5 October 2018, 70 (2019).

J. Surbrook, M. MacCormick, G. Bollen, M. Brodeur, M. Eibach, K. Gulyuz, A. Hamaker, C. Izzo, S. M. Lenzi, D. Puentes, M. Redshaw, R. Ringle, R. Sandler, S. Schwarz, P. Schury, N. Smirnova, C. Sumithrarachchi, **A. A. Valverde**, A. C. C. Villari, and I. T. Yandow, "Precision mass measurements of 44V and 44mV for nucleon-nucleon interaction studies", Hyperfine Interact. **240**, Proceedings of the 7th International Conference on Trapped Charged Particles and Fundamental Physics (TCP 2018), Traverse City, Michigan, USA, 30 September–5 October 2018, 65 (2019).

A. A. Valverde, M. Brodeur, D. P. Burdette, J. A. Clark, J. W. Klimes, D. Lascar, P. D. O'Malley, R. Ringle, G. Savard, and V. Varentsov, "Stopped, bunched beams for the TwinSol facility", Hyperfine Interact. **240**, Proceedings of the 7th International Conference on Trapped Charged Particles and Fundamental Physics (TCP 2018), Traverse City, Michigan, USA, 30 September–5 October 2018, 38 (2019).

D. P. Burdette, M. Brodeur, T. Ahn, J. Allen, D. W. Bardayan, F. D. Becchetti, D. Blankstein, G. Brown, B. Frentz, M. R. Hall, S. King, J. J. Kolata, J. Long, K. T. Macon, A. Nelson, P. D. O'Malley, C. Seymour, M. Skulski, S. Y. Strauss, and **A. A. Valverde**, "Resolving the discrepancy in the half-life of ^{20}F", Phys. Rev. C **99**, 015501 (2019).

W.-J. Ong, **A. A. Valverde**, M. Brodeur, G. Bollen, M. Eibach, K. Gulyuz, A. Hamaker, C. Izzo, D. Puentes, M. Redshaw, R. Ringle, R. Sandler, S. Schwarz, C. S. Sumithrarachchi, J. Surbrook, A. C. C. Villari, and I. T. Yandow, "Mass measurement of ^{51}Fe for the determination of the ^{51}Fe$(p, \gamma)^{52}$Co reaction rate", Phys. Rev. C **98**, 065803 (2018).

A. A. Valverde, M. Brodeur, T. Ahn, J. Allen, D. W. Bardayan, F. D. Becchetti, D. Blankstein, G. Brown, D. P. Burdette, B. Frentz, G. Gilardy, M. R. Hall, S. King, J. J. Kolata, J. Long, K. T. Macon, A. Nelson, P. D. O'Malley, M. Skulski, S. Y. Strauss, and B. Vande Kolk, "Precision half-life measurement of ^{11}C: The most precise mirror transition $\mathcal{F}t$ value", Phys. Rev. C **97**, 035503 (2018).

C. Izzo, G. Bollen, M. Brodeur, M. Eibach, K. Gulyuz, J. D. Holt, J. M. Kelly, M. Redshaw, R. Ringle, R. Sandler, S. Schwarz, S. R. Stroberg, C. S. Sumithrarachchi, **A. A. Valverde**, and A. C. C. Villari, "Precision mass measurements of neutron-rich Co isotopes beyond $N = 40$", Phys. Rev. C **97**, 014309 (2018).

A. A. Valverde, M. Brodeur, G. Bollen, M. Eibach, K. Gulyuz, A. Hamaker, C. Izzo, W.-J. Ong, D. Puentes, M. Redshaw, R. Ringle, R. Sandler, S. Schwarz, C. S. Sumithrarachchi, J. Surbrook, A. C. C. Villari, and I. T. Yandow, "High-Precision Mass Measurement of ^{56}Cu and the Redirection of the rp-Process Flow", Phys. Rev. Lett. **120**, 032701 (2018).

R. M. E. B. Kandegedara, G. Bollen, M. Eibach, N. D. Gamage, K. Gulyuz, C. Izzo, M. Redshaw, R. Ringle, R. Sandler, and **A. A. Valverde**, "β-decay Q values among the $A = 50$ Ti-V-Cr isobaric triplet and atomic masses of ^{46,47,49,50}Ti, ^{50,51}V, and $^{50,52-54}$Cr", Phys. Rev. C **96**, 044321 (2017).

J. Long, T. Ahn, J. Allen, D. W. Bardayan, F. D. Becchetti, D. Blankstein, M. Brodeur, D. Burdette, B. Frentz, M. R. Hall, J. M. Kelly, J. J. Kolata, P. D. O'Malley, B. E. Schultz, S. Y. Strauss, and **A. A. Valverde**, "Precision half-life measurement of ^{25}Al", Phys. Rev. C **96**, 015502 (2017).

C. Izzo, G. Bollen, S. Bustabad, M. Eibach, K. Gulyuz, D. Morrissey, M. Redshaw, R. Ringle, R. Sandler, S. Schwarz, and **A. A. Valverde**, "A laser ablation source for offline ion production at LEBIT", Nucl. Instrum. Methods B **376**, Proceedings of the XVIIth International Conference on Electromagnetic Isotope Separators and Related Topics (EMIS2015), Grand Rapids, MI, U.S.A., 11-15 May 2015, 60–63 (2016).

N. D. Gamage, G. Bollen, M. Eibach, K. Gulyuz, C. Izzo, R. M. E. B. Kandegedara, M. Redshaw, R. Ringle, R. Sandler, and **A. A. Valverde**, "Precise determination of the ^{113}Cd fourth-forbidden non-unique β-decay Q value", Phys. Rev. C **94**, 025505 (2016).

M. Eibach, G. Bollen, K. Gulyuz, C. Izzo, M. Redshaw, R. Ringle, R. Sandler, and **A. A. Valverde**, "Double resonant enhancement in the neutrinoless double-electron capture of ^{190}Pt", Phys. Rev. C **94**, 015502 (2016).

K. Gulyuz, G. Bollen, M. Brodeur, R. A. Bryce, K. Cooper, M. Eibach, C. Izzo, E. Kwan, K. Manukyan, D. J. Morrissey, O. Naviliat-Cuncic, M. Redshaw, R. Ringle, R. Sandler, S. Schwarz, C. S. Sumithrarachchi, **A. A. Valverde**, and A. C. C. Villari, "High Precision Determination of the β Decay Q_{EC} Value of ^{11}C and Implications on the Tests of the Standard Model", Phys. Rev. Lett. **116**, 012501 (2016).

D. Lincoln, R. Baker, A. Benjamin, G. Bollen, M. Redshaw, R. Ringle, S. Schwarz, A. Sonea, and **A. A. Valverde**, "Development of a high-precision Penning trap magnetometer for the LEBIT facility", Int. J. of Mass Spectrom. **379**, 1–8 (2015).

M. Redshaw, A. L. Benjamin, G. Bollen, R. Ferrer, D. L. Lincoln, R. Ringle, S. Schwarz, and **A. A. Valverde**, "Fabrication and characterization of field emission points for ion production in Penning trap applications", Int. J. of Mass Spectrom. **379**, 187–193 (2015).

M. Eibach, G. Bollen, M. Brodeur, K. Cooper, K. Gulyuz, C. Izzo, D. J. Morrissey, M. Redshaw, R. Ringle, R. Sandler, S. Schwarz, C. S. Sumithrarachchi, **A. A. Valverde**, and A. C. C. Villari, "Determination of the Q_{EC} values of the $T = 1/2$ mirror nuclei ^{21}Na and ^{29}P at LEBIT", Phys. Rev. C **92**, 045502 (2015).

A. A. Valverde, G. Bollen, M. Brodeur, R. A. Bryce, K. Cooper, M. Eibach, K. Gulyuz, C. Izzo, D. J. Morrissey, M. Redshaw, R. Ringle, R. Sandler, S. Schwarz, C. S. Sumithrarachchi, and A. C. C. Villari, "First Direct Determination of the Superallowed β-Decay Q_{EC} Value for ^{14}O", Phys. Rev. Lett. **114**, 232502 (2015).

K. Gulyuz, J. Ariche, G. Bollen, S. Bustabad, M. Eibach, C. Izzo, S. J. Novario, M. Redshaw, R. Ringle, R. Sandler, S. Schwarz, and **A. A. Valverde**, "Determination of the direct double-β-decay Q value of ^{96}Zr and atomic masses of $^{90-92,94,96}$Zr and $^{92,94-98,100}$Mo", Phys. Rev. C **91**, 055501 (2015).

A. A. Valverde, G. Bollen, K. Cooper, M. Eibach, K. Gulyuz, C. Izzo, D. J. Morrissey, R. Ringle, R. Sandler, S. Schwarz, C. S. Sumithrarachchi, and A. C. C. Villari, "Penning trap mass measurement of ^{72}Br", Phys. Rev. C **91**, 037301 (2015).

S. Bustabad, G. Bollen, M. Brodeur, D. L. Lincoln, S. J. Novario, M. Redshaw, R. Ringle, S. Schwarz, and **A. A. Valverde**, "First direct determination of the ^{48}Ca double-β decay Q value", Phys. Rev. C **88**, 022501 (2013).

Invited Talks

7/19/2019   **The $N = 126$ Beam Factory at ANL**, *The 13th International Conference on Stopping and Manipulation of Ions and Related Topics (SMI-2019)*, Montreal, QC, Canada.

11/2/2018   **Precision Mass Measurements for Explosive Nuclear Astrophysics**, *Heavy Ion Discussion*, Physics Division, Argonne National Laboratory, Lemont, IL.

10/9/2018   **Precision Mass Measurements for Astrophysical Applications at LEBIT and the CPT**, *Nuclear Physics Seminar*, TRIUMF, Vancouver, BC, Canada.

■■■■ Contributed Talks

8/8/2019   **Canadian Penning Trap Update**, *2019 Low Energy Community Meeting*, Durham, NC.

8/10/2018   **Precision Measurements at the University of Notre Dame**, *2018 Low Energy Community Meeting*, East Lansing, MI.

5/21/2018   **Penning Trap Mass Measurement of ^{56}Cu and the redirection of the rp-process flow**, *2018 JINA-CEE Frontiers in Nuclear Astrophysics Junior Researcher Workshop*, Notre Dame, IN.

10/27/2017   **Penning Trap Mass Measurement of ^{56}Cu**, *2017 Fall Meeting of the APS Division of Nuclear Physics*, Pittsburgh, PA.

■■■■ Poster Presentations

10/2/2018   **Stopped, Bunched Beams for the *TwinSol* Facility**, *7th International Conference on Trapped Charged Particles and Fundamental Physics 2018 (TCP 2018)*, Traverse City, MI.

9/18/2018   **An RFQ Cooler-Buncher for the $N = 126$ factory at Argonne National Laboratory**, *EMIS 2018 – 18th International Converence on Electromagnetic Isotope Separators and Related Topics*, Geneva, GE, CH.

5/23/2018   **Penning Trap Mass Measurement of ^{56}Cu and the redirection of the rp-process flow**, *2018 JINA-CEE Frontiers in Nuclear Astrophysics Meeting*, Notre Dame, IN.

5/12/2015   **SWIFT Beam Purification at LEBIT**, *EMIS-2015 – 17th International Conference on Electromagnetic Isotope Separators and Related Topics*, Grand Rapids, MI.

8/2/2013   **Recent Developments at the LEBIT Facility at NSCL/FRIB *and* SIPT—An Ultrasensitive Mass Spectrometer for Rare Isotopes**, *Exotic Beam Summer School 2013*, Berkeley, CA.

6/5/2013   **SIPT—An Ultrasensitive Mass Spectrometer for Rare Isotopes**, *UITI2013*, Lansing, MI.

Printed in the United States
By Bookmasters